student edition

Critical Thinking for Life!
Mentoring Minds

Publisher
Michael L. Lujan, M.Ed.

Editorial Director
Teresa Sherman, B.S.E.

Production Coordinator
Kim Barnes, B.B.A.

Digital Production Artists
Ashley E. Francis, A.A.
Tammy Black

Illustrators
Gabe Urbina, A.A.S.
Judy Bankhead, M.F.A.
Fernando C. Salinas

Content Development Team
Marian Rainwater, M.Ed.
Amanda Byers, B.S.
Paula Jones, M.S.
Rebecca Clements, B.S.
Heather Seib, B.S.
Mindy Rucker, M.S.
Dana White, B.S.E.
Leslie Huval, B.A.
Casandra Hartsfield, M.Ed.
Cathy K. Smith, M.Ed.

Content Editorial Team
Allison Wiley, B.S.E.
Marian Rainwater, M.Ed.
Amanda Byers, B.S.
Paula Jones, M.S.
Jennifer Mallios, B.A.
Nancy Roseman, B.S.E.
Patricia Martin, M.Ed.
Amy L. Norris, M.Ed.
Cathy Cutler, B.S.E.
Diane Sorrels, B.A.
Nan Hutchins Bailey, M.S.

Critical Thinking for Life!™
Mentoring Minds

PO Box 8843 • Tyler, TX 75711

[p] 800.585.5258 • [f] 800.838.8186

For other great products from Mentoring Minds,
please visit our website at:
mentoring**minds**.com

ISBN: 978-1-938935-67-1

Motivation Math™, Student Edition

Copyright infringement is a violation of Federal Law.

Trademarks used in this book belong to their respective owners.

©2014 by **Mentoring Minds, L.P.** All rights reserved. *Motivation Math*™ is protected by copyright law and international treaties. No part of this publication may be reproduced, translated, stored in a retrieval system, or transmitted in any way or by any means (electronic, mechanical, photocopying, recording, or otherwise) without prior written permission from *Mentoring Minds, L.P.* Unauthorized reproduction of any portion of this book without written permission of *Mentoring Minds, L.P.* may result in criminal penalties and will be prosecuted to the maximum extent of the law. Not for resale.

Printed in the United States of America

motivationmath™
Table of Contents

Unit 1 – Classify whole numbers, integers, and rational numbers 6.2(A)–S 7

Unit 2 – Identify a number, its opposite, and its absolute value 6.2(B)–S 15

Unit 3 – Locate, compare, and order integers and rational numbers using a number line 6.2(C)–S 23

Unit 4 – Order rational numbers 6.2(D)–R 31

Unit 5 – Interpret fractions as division 6.2(E)–S 39

Unit 6 – Divide by a rational number and multiply by its reciprocal 6.3(A)–S 47

Unit 7 – Determine whether a quantity is increased or decreased when multiplied by a fraction 6.3(B)–S 55

Unit 8 – Represent integer operations with concrete models 6.3(C)–S 63

Unit 9 – Add, subtract, multiply, and divide integers fluently 6.3(D)–R 71

Unit 10 – Multiply and divide positive rational numbers fluently 6.3(E)–R 79

Unit 11 – Differentiate between additive and multiplicative relationships 6.4(A)–S 87

Unit 12 – Apply qualitative and quantitative reasoning to solve problems involving ratios and rates 6.4(B)–R 95

Unit 13 – Give examples of ratios as multiplicative comparisons of two quantities 6.4(C)–S 103

Unit 14 – Give examples of rates as the comparison by division of two quantities 6.4(D)–S 111

Unit 15 – Represent ratios, percents, and benchmark fractions using models, fractions, and decimals 6.4(E)–S, 6.4(F)–S 119

Unit 16 – Generate equivalent forms of fractions, decimals, and percents 6.4(G)–R 127

Unit 17 – Convert units within a measurement system 6.4(H)–R 135

Unit 18 – Represent mathematical and real-world problems involving ratios and rates 6.5(A)–S 143

Unit 19 – Solve real-world problems involving percents 6.5(B)–R 151

Unit 20 – Use equivalent fractions, decimals, and percents to show equal parts of the same whole 6.5(C)–S 159

Unit 21 – Identify and write equations for independent and dependent quantities 6.6(A)–S, 6.6(B)–S 167

Unit 22 – Represent a situation in the form of $y = kx$ or $y = x + b$ 6.6(C)–R 175

Unit 23 – Generate equivalent numerical expressions using order of operations 6.7(A)–R 183

Unit 24 – Distinguish between expressions and equations; determine if two expressions are equivalent 6.7(B)–S, 6.7(C)–S 191

Unit 25 – Generate equivalent expressions using the properties of operations 6.7(D)–R 199

Unit 26 – Extend previous knowledge of triangles and their properties 6.8(A)–S 207

Unit 27 – Model area formulas by decomposing and rearranging parts 6.8(B)–S 215

Unit 28 – Write equations and determine solutions for problems involving area and volume 6.8(C)–S, 6.8(D)–R 223

Unit 29 – Write one-variable, one-step equations and inequalities and represent their solutions 6.9(A)–S, 6.9(B)–S 231

Unit 30 – Write corresponding real-world problems given one-variable, one-step equations, or inequalities 6.9(C)–S 239

Unit 31 – Model and solve one-variable, one-step equations, and inequalities 6.10(A)–R 247

Unit 32 – Determine if the given values make one-variable, one-step equations, or inequalities true 6.10(B)–S 255

Unit 33 – Graph points in all four quadrants 6.11(A)–R 263

Unit 34 – Represent numeric data graphically 6.12(A)–S 271

Unit 35 – Use the graphical representation of numeric data to describe the center, spread, and shape of the data distribution 6.12(B)–S 279

Unit 36 – Summarize numerical data with numerical summaries and use these summaries to describe the center, spread, and shape of the data distribution 6.12(C)–R 287

Unit 37 – Summarize categorical data with numerical and graphical summaries and use these summaries to describe the data distribution 6.12(D)–R 295

Unit 38 – Interpret summaries of numeric data 6.13(A)–R 303

Unit 39 – Distinguish between situations that yield data with and without variability 6.13(B)–S 311

Unit 40 – Compare the features and cost of a checking account and a debit card; balance a check register 6.14(A)–S, 6.14(C)–S 321

Unit 41 – Distinguish between debit cards and credit cards; explain why it is important to establish a positive credit history; describe the information in a credit report and the value of a credit report 6.14(B)–S, 6.14(D), 6.14(E)–S, 6.14(F)–S 327

Unit 42 – Explain various methods to pay for college; compare the annual salary of several occupations and calculate the effects on lifetime income 6.14(G)–S, 6.14(H)–S 333

Performance Assessments 339

Chart Your Success 357

Math Glossary 361

Grade 6 Mathematics Reference Materials 383

Name _____

Standard 6.2(A) – Supporting

Unit 1 Introduction

1 Study the diagram below.

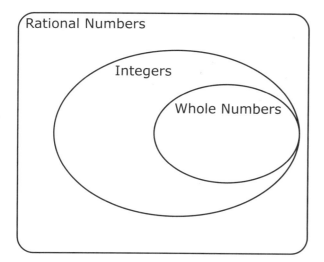

For the numbers listed, determine the set or sets to which each belongs.

-2 _____

$\frac{3}{4}$ _____

-6,842 _____

292 _____

$-\frac{19}{3}$ _____

2 Refer to the diagram above to complete the sentence:

The set of _____ is a subset of the set of _____, which is a subset of the set of _____.

3 List the numbers that fit each description.

Whole numbers less than 5

_____

Integers between -4 and 2, inclusive

_____

Rational numbers with a denominator of 4 that fall between -1 and -2, exclusive

_____

4 Write each number in the correct region on the diagram below.

-109 $\frac{7}{9}$ $3.\overline{3}$

-8.4   1   -38

0   25.5 $\frac{16}{4}$

0.003   248   -16

```
┌─────────────────────────────────┐
│  Rational Numbers               │
│   ┌──────────────────────────┐  │
│   │  Integers                │  │
│   │   ┌──────────────────┐   │  │
│   │   │  Whole Numbers   │   │  │
│   │   │                  │   │  │
│   │   │                  │   │  │
│   │   └──────────────────┘   │  │
│   └──────────────────────────┘  │
└─────────────────────────────────┘
```

5 Enrique writes the following incorrect statement:

All rational numbers are also whole numbers.

Rewrite Enrique's statement to make it true.

_____

_____

6 List three numbers that fit into each of the following number sets.

Whole numbers _____

Integers _____

Rational numbers _____

Unit 1 Guided Practice

Standard 6.2(A) – Supporting

1. Which diagram below correctly displays the relationship between sets of numbers?

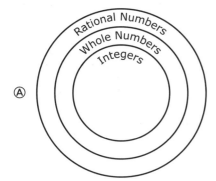
Ⓐ

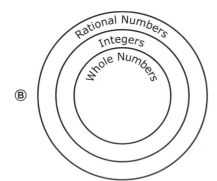
Ⓑ

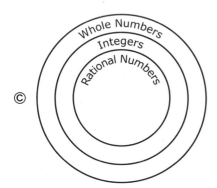
Ⓒ

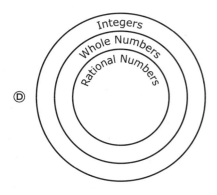
Ⓓ

2. Which of the following statements is always true?

   Ⓕ A rational number is a whole number.

   Ⓖ A whole number is not a rational number.

   Ⓗ Whole numbers are also integers and rational numbers.

   Ⓙ A rational number is an integer.

3. Which best describes the classification for the following number set?

   $\{3, 6, 12, 35, 56\}$

   I. Rational numbers
   II. Integers
   III. Whole numbers

   Ⓐ I only

   Ⓑ III only

   Ⓒ II and III only

   Ⓓ I, II, and III

4. Given the set of numbers $\{10, -\frac{5}{4}, 1.25, -2\}$, which are classified as integers?

   Ⓕ $10, -\frac{5}{4},$ and -2 only

   Ⓖ 10 and -2 only

   Ⓗ $-\frac{5}{4}$ and -2 only

   Ⓙ $-\frac{5}{4}$ and 1.25 only

Standard 6.2(A) — Supporting

Unit 1 Independent Practice

1. Which best describes the classification for the following number set?

$$\{-4\tfrac{1}{2},\ -1,\ \tfrac{3}{4},\ 3.5,\ 13,\ 24\tfrac{5}{8}\}$$

   I. Rational numbers
   II. Integers
   III. Whole numbers

   Ⓐ I only
   Ⓑ I and II only
   Ⓒ I and III only
   Ⓓ I, II, and III

2. Which lists the correct labels for the diagram shown?

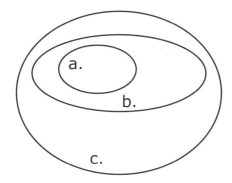

   Ⓕ a. Rational Numbers
      b. Whole Numbers
      c. Integers

   Ⓖ a. Whole Numbers
      b. Rational Numbers
      c. Integers

   Ⓗ a. Rational Numbers
      b. Integers
      c. Whole Numbers

   Ⓙ a. Whole Numbers
      b. Integers
      c. Rational Numbers

3. During a lesson in math class, Deja was asked by her teacher to determine where −81 would lie in a Venn diagram. Which of the illustrated Venn diagrams shows the best placement of −81?

Ⓐ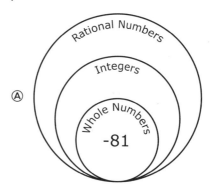

Ⓑ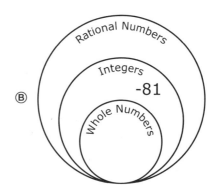

Ⓒ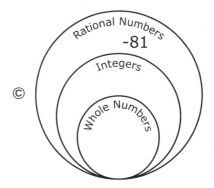

Ⓓ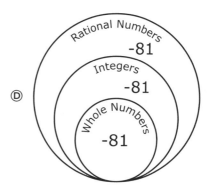

Unit 1 Assessment

Name _____

Standard 6.2(A) – Supporting

1. Which of the following hierarchy diagrams could best be used to demonstrate the relationship between sets of numbers?

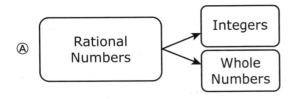

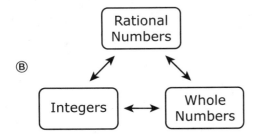

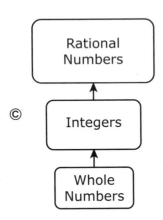

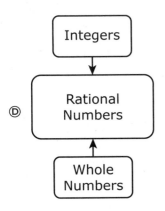

2. The set of rational numbers includes all the following except —

Ⓕ −42

Ⓖ 14.7

Ⓗ 0

Ⓙ $\frac{5}{0}$

3. Study the diagram below.

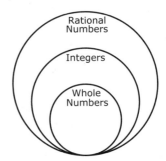

In which of the following sets can each number only be classified as rational?

Ⓐ $\{0, -\frac{18}{7}, 124, -3, \frac{2}{3}\}$

Ⓑ $\{3.5, -4, 23.8, -1.2, 0\}$

Ⓒ $\{-\frac{5}{8}, \frac{13}{2}, 23\frac{2}{3}, -\frac{41}{49}, \frac{8}{5}\}$

Ⓓ $\{-7.75, -5.8, 0.125, 4, 21\}$

4. Which of the following statements is always true?

Ⓕ All integers are whole numbers.

Ⓖ All whole numbers are integers.

Ⓗ Some whole numbers are not rational numbers.

Ⓙ All rational numbers are also integers.

Name _____

Standard 6.2(A) – Supporting

Unit 1 Critical Thinking

Janson draws the following diagram to use in classifying numbers.

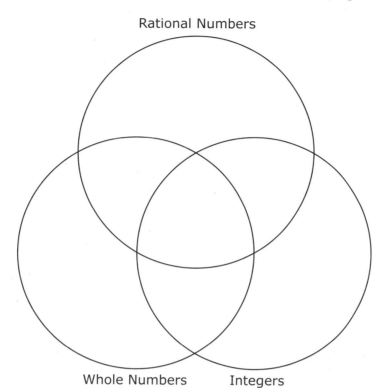

Is Janson's diagram the best choice for classifying numbers? Justify your answer.

_____
_____
_____
_____

Are there any suggestions you would give Janson to improve his diagram? If so, sketch your ideas below. If not, explain your reasoning.

Unit 1 Journal/Vocabulary Activity

Standard 6.2(A) – Supporting

Journal

Mrs. Anderson writes the following statement on the board: *All whole numbers are also integers, but all integers are not whole numbers.* Explain what Mrs. Anderson means by this statement. Is her statement always true? Explain why or why not.

_____

_____

_____

_____

Vocabulary Activity

Use the clues to unscramble the vocabulary words below. Then, unscramble the circled letters to reveal Johnny's response to his teacher.

| A I L A N R O T | B N M E U R | Can be written as $\frac{a}{b}$ as long as $b \neq 0$ |

| T G E I N R E S | | The set of whole numbers and their opposites |

| E W L O H | B N E U M R S | Numbers used to count and 0 |

| I N T T A E N I G R M | I C L M E D A | Number with a fixed number of digits following the decimal point |

| P R G T E E N I A | M D A L I E C | Number with a sequence of digits repeating infinitely following the decimal point |

| E S T | | Collection of distinct elements |

| U B E T S S | | A set within a set |

| T P O S I E I V | U M B E R N | A number greater than zero |

| T V G I E A E N | B R U M N E | A number less than zero |

Johnny: I used to hate math!
Teacher: What changed your mind?
Johnny: ◯ ◯E◯◯◯Z◯◯ ◯◯C◯◯◯L◯ ◯◯V◯ A ◯◯I◯◯ !

Bonus Race

Play *Bonus Race* with a partner. Each pair of players needs a game board, one die, and two game tokens. Player 1 rolls the die and moves the game token the number of spaces indicated. Player 1 has the opportunity for one bonus move according to the following rules:

- If the space contains a number that can be classified only as a rational number, move ahead one space.
- If the space contains a number that can be classified only as a rational number *and* an integer, move ahead two spaces.
- If the space contains a number that can be classified as a rational number, an integer, *and* a whole number, move ahead three spaces.

Player 2 repeats the process. The first player to reach or pass the *Finish* space is the winner.

Start	$\frac{3}{4}$	-47	$\frac{1}{2}$	2	$\frac{3}{10}$	0.75	$\frac{8}{4}$	$-\frac{2}{3}$	-0.9	0
										$\frac{21}{7}$
	12	$3\frac{1}{2}$	-15	$\frac{2}{5}$	0.1	$\frac{16}{4}$	$-\frac{5}{6}$	44		$2\frac{1}{2}$
	$-\frac{9}{3}$							$\frac{35}{5}$		-1
	-0.6	-8	$\frac{7}{8}$	32	$\frac{10}{9}$	0.3		-13		0.4
						Finish		$-\frac{8}{3}$		$-\frac{6}{3}$
								7		9
	-6	$\frac{42}{5}$	14	$-\frac{17}{2}$	0.62	$\frac{3}{8}$	-23	$\frac{12}{3}$	0.15	-0.88
	0.8									$\frac{1}{8}$
	$\frac{4}{9}$	-34	$-\frac{24}{5}$	$\frac{15}{3}$	-1.4	23	$\frac{3}{2}$	$-\frac{18}{2}$	-10	$-\frac{10}{5}$	0.25

Unit 1 Homework

Standard 6.2(A) – Supporting

1. In your own words, write a definition for the following terms.

   Rational number: _____

   _____

   Integer: _____

   _____

   Whole Number: _____

   _____

   Draw a diagram illustrating the relationship between the terms. Include two numbers in each region of your diagram.

2. Susan and James each believe they have written a number that represents both a rational number and an integer. Susan's number is $-\frac{2}{3}$. James' number is -4. Which student is correct? Explain your answer.

   _____
   _____
   _____
   _____

3. A diagram showing the relationship between number sets is shown below.

   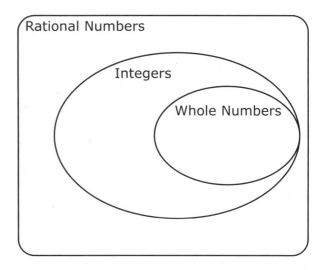

   Write three numbers in each region of the diagram.

4. For each number listed below, identify the set or sets of numbers to which each belongs.

   $-\frac{8}{5}$ _____

   16 _____

   -68 _____

   $-\frac{12}{3}$ _____

   37.92 _____

5. Explain why $\frac{2}{0}$ is not a rational number.

   _____
   _____
   _____

Connections

Construct a target on a sheet of paper by drawing one large circle (as large as the paper), with two additional circles, one inside the other. Label the circles from inner to outer with the terms *whole numbers*, *integers*, and *rational numbers*. Lay the target on the floor and drop a penny over the target. Name a number that is a member of the subset in which the penny landed. Have a parent write the numbers on the target where the penny landed.

Name _____

Standard 6.2(B) – Supporting

Unit 2 Introduction

Use the number line to answer the following questions.

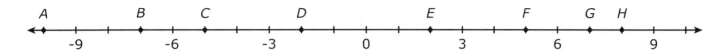

1. What value does point E represent on the number line?
   _____

2. Which point represents the opposite of point G?
   _____
   Explain how you know.
   _____
   _____

3. What is the absolute value of point A?
   _____

4. Which pairs of points have the same absolute value?
   _____
   _____
   _____

5. Write the following values.
   F _____
   Opposite of F _____
   Absolute value of F _____

6. Which point best represents a withdrawal of $10?
   _____

7. Which point represents the opposite of a decrease of 5 degrees?
   _____
   Explain your answer.
   _____
   _____

8. How can you determine the absolute value of ⁻7 using a number line?
   _____
   _____
   _____

Unit 2 Guided Practice

Standard 6.2(B) – Supporting

1 A great white shark can swim 50 feet below sea level. Which best represents the opposite of the depth the great white shark can swim?

Ⓐ -50 feet

Ⓑ -|-50| feet

Ⓒ -|50| feet

Ⓓ 50 feet

2 Which equation represents the absolute value of -2?

Ⓕ |-2| = -2

Ⓖ |2| = -2

Ⓗ |-2| = 2

Ⓙ -|2| = 2

3 Calvin states that the absolute values of opposite numbers have the same value. Enrique disagrees with Calvin and says the absolute values of opposite numbers are still opposite numbers. Who is correct?

Ⓐ Calvin, because 19 and -19 are opposites, and |19| = -19 while |-19| = -19

Ⓑ Enrique, because 19 and -19 are opposites, while |19| = -19 and |-19| = 19

Ⓒ Calvin, because 19 and -19 are opposites, while |19| = 19 and |-19| = 19

Ⓓ Cannot determine because it depends on which numbers are being discussed

4 Which model shows the opposite of -6?

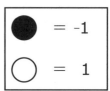

Ⓕ

Ⓖ

Ⓗ

Ⓙ

5 Which statement is NOT true about the number line shown?

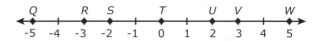

Ⓐ The opposite of point R is equivalent to the value of point V.

Ⓑ The absolute value of point S is the same as the value of point S.

Ⓒ The opposite of point T is equivalent to the value of point T.

Ⓓ The absolute value of point Q is the same as the absolute value of point W.

6 Montel needs to determine the distance from -3 to zero. Which statement best describes the process Montel should use?

Ⓕ Find the absolute value of negative three.

Ⓖ Find the opposite of positive three.

Ⓗ Find the absolute value of positive three.

Ⓙ Find the opposite of negative three.

Name _____

Standard 6.2(B) – Supporting

Unit 2 Independent Practice

1. Lamont lost 8 yards in the first quarter of the football game. In the second quarter, he gained 6 yards. Which integer represents the opposite of Lamont's yardage in the second quarter?

   Ⓐ -8
   Ⓑ -6
   Ⓒ 6
   Ⓓ 8

2. If $|g| = 18$, which values for g are true?

   I. 18
   II. 0
   III. -18
   IV. 81

   Ⓕ I only
   Ⓖ I, II, and III only
   Ⓗ I and III only
   Ⓙ I, III, and IV only

3. Jeremiah recorded the temperatures for the week in his science project journal. From his data, Jeremiah noticed that the high temperature dropped 15 degrees from Tuesday to Wednesday. Which best represents Jeremiah's observation?

   Ⓐ 15.0
   Ⓑ $|-15|$
   Ⓒ -15
   Ⓓ $|15|$

4. What integer is the opposite of the value of point V on the number line?

   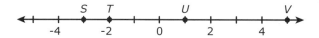

   Record your answer and fill in the bubbles on the grid below. Be sure to use the correct place value.

   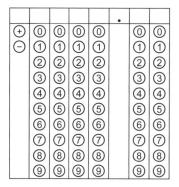

5. Mrs. Martin is thinking of two mystery numbers. She gives two clues as shown in the table.

   Mystery Numbers

Clue 1:	The numbers are opposites.
Clue 2:	The distance between the numbers on the number line is 20.

   What are the mystery numbers?

   Ⓐ -5 and 5
   Ⓑ -10 and 10
   Ⓒ -15 and 15
   Ⓓ -20 and 20

6. The low Fahrenheit temperature on Tuesday was 9 degrees below zero. The high temperature on Wednesday was 0°F. What is the absolute value of the low temperature on Tuesday?

   Ⓕ $|-9| = -9$
   Ⓖ $|9| = -9$
   Ⓗ $|9| = 9$
   Ⓙ $|-9| = 9$

Unit 2 Assessment

Standard 6.2(B) – Supporting

Use the number line to answer questions 1 and 2.

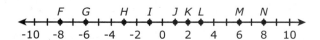

1. Which statement is NOT true about the number line?

   Ⓐ Points *L* and *H* represent opposite integers.

   Ⓑ The value of point *G* is equivalent to the absolute value of point *M*.

   Ⓒ The value of point *F* is the opposite of the value of point *N*.

   Ⓓ The absolute value of point *I* is the same as the absolute value of point *J*.

2. Which integer is the opposite of the value of point *K*?

   Ⓕ 3

   Ⓖ 2

   Ⓗ -3

   Ⓙ -2

3. Which of the following equations is true?

   I. |25| = 25
   II. |-56| = -56
   III. |-38| = 38
   IV. |89| = -89

   Ⓐ I only

   Ⓑ II and IV only

   Ⓒ I and III only

   Ⓓ I, II, and III only

4. The temperature was 8 degrees below zero on Monday. On Tuesday, the temperature was 1 degree above zero. Which best represents the opposite of Tuesday's temperature?

   Ⓕ |1|

   Ⓖ -1

   Ⓗ |-1|

   Ⓙ 1

5. Jakavian locates 9 on the number line. She determines it is 9 units from zero. If Jakavian needs to find a different number that is also 9 units from zero, what does she need to do?

   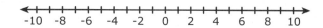

   Ⓐ Find the absolute value of 9, because the absolute value of a number is the number's distance from zero.

   Ⓑ Find the opposite of -9, because opposite numbers are an equal distance from zero.

   Ⓒ Find the opposite of 9, because opposite numbers are an equal distance from zero.

   Ⓓ There is not a different number that is also 9 units from zero.

6. Tony spent $15 buying music on iTunes®. Afterwards, his dad added $20 to Tony's iTunes® account. Which integer best represents the transaction on Tony's iTunes® account before his dad's deposit?

   Ⓕ 20

   Ⓖ 15

   Ⓗ -15

   Ⓙ -20

Standard 6.2(B) – Supporting

Unit 2 Critical Thinking

1 When students entered Mr. Wong's classroom, they noticed a large number line on the floor with zero marked in the center. Mr. Wong gave each student an integer written on a sheet of paper. He asked two students to stand at their integers on the number line without displaying their papers to the class. Mr. Wong asked the class if the locations of the two students represented opposite numbers. Explain how you would answer Mr. Wong's question.

_____
_____
_____
_____

2 Determine the value of x for the equation $|x| = -\frac{1}{2}$. Explain your answer.

_____
_____
_____
_____

3 Two students illustrate the definition of absolute value by graphing $|-8|$ on a number line.

Student A
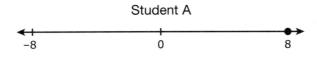

Student B
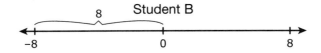

Which student's illustration is correct? Explain your answer.

_____
_____
_____

Unit 2 Journal/Vocabulary Activity

Name _____

Standard 6.2(B) – Supporting

Journal

What distance is traveled on an elevator going up 2 floors? _____

What distance is traveled on an elevator going down 2 floors? _____

Explain how the distance an elevator travels is like absolute value.

_____
_____
_____

Vocabulary Activity

Create a concept web for the following vocabulary words:

zero, positive integers, negative integers, opposite numbers, absolute value

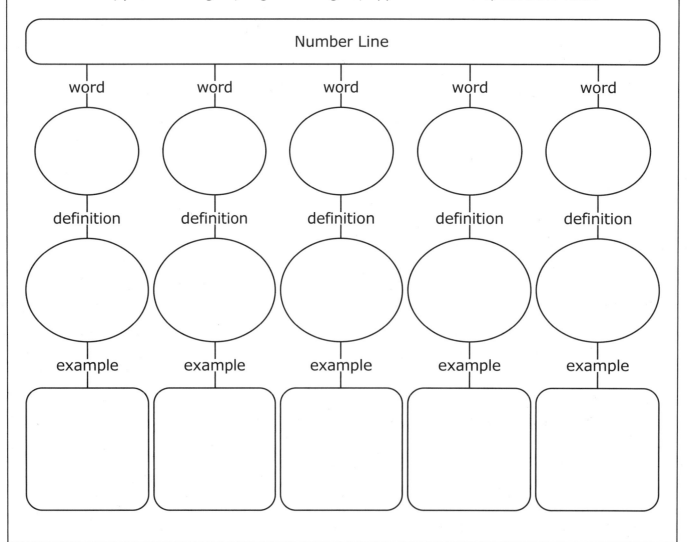

Name _____

Standard 6.2(B) — Supporting

Unit 2 Motivation Station

Opposites Memory

Play *Opposites Memory* with a partner. Each pair of players needs a game board and 36 Color Tiles® or 1-inch squares. Players cover each square on the game board with a color tile. Player 1 removes two tiles. If the squares revealed show values that are opposites, the player keeps the tiles. If the squares revealed show values that are not opposites, the tiles are returned to the game board. Play passes to player 2, and the process is repeated. The game ends when all tiles are claimed. The player with more tiles is the winner.

$|10|$	13	$-|5|$	-1	-8	-3
-17	2	$-|12|$	-11	5	$-|9|$
$|-4|$	7	6	0	14	$|-12|$
-14	16	-6	-15	$-|10|$	$|-17|$
$|-8|$	0	-16	11	-4	-2
15	3	$|-9|$	-7	$|-1|$	$-|13|$

©2014 mentoringminds.com   motivationmath™ LEVEL 6   ILLEGAL TO COPY   21

Unit 2 Homework

Standard 6.2(B) – Supporting

1 Draw and label each point on the number line.

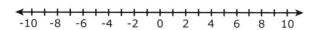

Point A: The opposite of 6

Point B: A number that has the same absolute value as ⁻8

Point C: A number that has the same absolute value as 3

Point D: The opposite of ⁻4

2 Ainsley made a $75 deposit to her savings account. What integer represents the deposit?

_____

What situation would describe the opposite of this transaction?

_____

What integer represents the opposite of Ainsley's transaction?

_____

3 Explain why the distance from 0 to ⁻7 is the same as the distance from 0 to 7.

_____

_____

4 Calvin modeled positive 8 using the key shown below. Draw the opposite of Calvin's model.

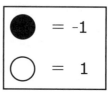

5 Determine all values that are true for the equation $|x| = 23$.

_____

6 Mrs. Peacock asked her students to find a number that has an absolute value of 6. Kevin thinks the answer is ⁻6, but Ryan argues that the answer is 6. Who is correct?

_____

Use the number line to explain why.

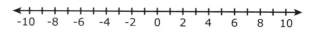

_____

_____

Connections

1. Remove the face cards from a deck of cards. Black cards represent positive numbers and red cards represent negative numbers. Flip over a card and tell the opposite and the absolute value of the number on the card.

2. Record 2 real-world examples of absolute value, like the elevator example used in the journal.

Name _____

Standard 6.2(C) – Supporting

Unit 3 Introduction

Use the number line to answer questions 1–3.

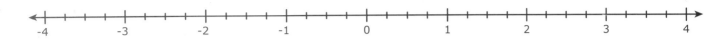

1. Arrange the following numbers in order from least to greatest.

   $3.5, -\frac{1}{4}, -2\frac{3}{4}, 0, 1$

   _____

2. Using the number line, place the following points in the appropriate position on the line. Label each point above the line.

   $A = \frac{1}{2}$

   $B = -1\frac{1}{4}$

   $C = -3\frac{3}{4}$

   $D = 2\frac{1}{2}$

   $E = 1\frac{1}{4}$

3. During a chemistry lab, Joel used an eye dropper to place chemicals in a beaker. For one element he needed to use less than $\frac{3}{4}$ milliliter but more than $\frac{1}{4}$ milliliter. List three rational numbers, in either fraction or decimal form, that could be the amount Joel used for his lab.

   _____

4. Marilla's teacher gave her four fractions to locate on a number line:

   $\frac{1}{3}, \frac{7}{10}, \frac{2}{3},$ and $\frac{5}{8}$

   Place the fractions in the appropriate location on the number line below.

   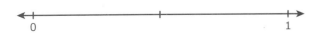

5. The local news station posted the following information about area lake levels on their website.

   Area Lake Levels

Joe Pool Lake	-0.71 ft
Lake Lewisville	-4.9 ft
Possum Kingdom Lake	-10.73 ft
Lake Ray Hubbard	-2.81 ft
Lake Whitney	-8.86 ft

   Place points on the number line to represent the level of each lake.

   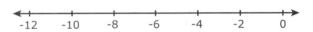

   List the lakes in order from most full to least full.

   _____
   _____
   _____
   _____

6. Use the number line below to compare ⁻2.4 and ⁻2.45. Place the correct symbol (>, <, or =) in the blank.

   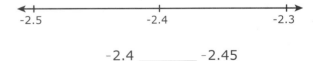

   ⁻2.4 _____ ⁻2.45

   Explain how you determined your answer.

   _____
   _____

Unit 3 Guided Practice

Standard 6.2(C) – Supporting

Use the number line below to answer questions 1 and 2.

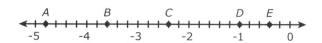

1. Which of the following points best corresponds to $-3\frac{3}{5}$?

    Ⓐ Point B

    Ⓑ Point D

    Ⓒ Point E

    Ⓓ Point C

2. Point E on the number line corresponds to which of the following values?

    Ⓕ $-1\frac{3}{5}$

    Ⓖ $-1\frac{1}{2}$

    Ⓗ $-\frac{2}{5}$

    Ⓙ $-\frac{1}{3}$

3. Based on the model below, which of the following is a true statement?

    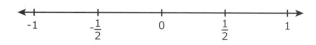

    Ⓐ $-\frac{5}{8} > -\frac{1}{2}$

    Ⓑ $0 < -0.1$

    Ⓒ $0.25 < -\frac{1}{4}$

    Ⓓ $-0.002 > -0.020$

4. Use the model below to answer the question that follows.

    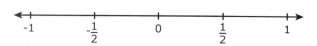

    Which of the following numbers is less than $-\frac{2}{3}$?

    Ⓕ $-\frac{1}{2}$

    Ⓖ $\frac{1}{3}$

    Ⓗ $-\frac{7}{8}$

    Ⓙ $-\frac{1}{6}$

Use the number line below to answer questions 5 and 6.

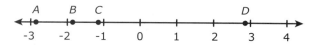

5. Which list shows rational numbers in order from greatest to least?

    Ⓐ $3, 1\frac{1}{2}, 1, -2\frac{1}{2}, -2, -\frac{1}{2}$

    Ⓑ $3, 1\frac{1}{2}, 1, -\frac{1}{2}, -2, -2\frac{1}{2}$

    Ⓒ $3, -2\frac{1}{2}, -2, 1\frac{1}{2}, 1, -\frac{1}{2}$

    Ⓓ $-2\frac{1}{2}, -2, -\frac{1}{2}, 1, 1\frac{1}{2}, 3$

6. Which of the following best describes the location of -2.8 on the number line?

    Ⓕ Point A

    Ⓖ Point B

    Ⓗ Point C

    Ⓙ Point D

Name _____

Standard 6.2(C) — Supporting

Unit 3 Independent Practice

1. Hannah collected water samples to compare amounts of rainfall. Which measurement is closest to 2 inches of rainfall?

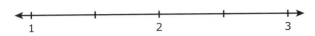

Ⓐ $\frac{3}{2}$ in.

Ⓑ 2.25 in.

Ⓒ $1\frac{7}{8}$ in.

Ⓓ 2.2 in.

2. Which of the following number lines is labeled correctly?

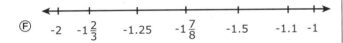

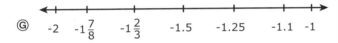

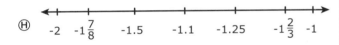

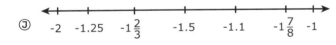

3. Nyasha cut four ribbons to use for her hair. She cut a blue ribbon $2\frac{7}{8}$ inches long, a red ribbon 2.85 inches long, a yellow ribbon $\frac{9}{4}$ inches long, and a green ribbon $\frac{7}{3}$ inches long. Which ribbon was the shortest?

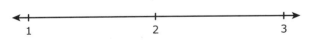

Ⓐ Blue   Ⓒ Green

Ⓑ Red   Ⓓ Yellow

Use the number line below to answer questions 4 and 5.

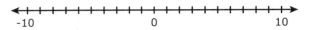

4. The temperatures in five Connecticut cities on one day in January are -7°F, -3°F, 5°F, -4°F, and -6°F. The median temperature is the temperature in the middle of the group when the numbers are ordered. What is the median temperature for the five cities?

Ⓕ -4°F

Ⓖ 5°F

Ⓗ -6°F

Ⓙ -3°F

5. Which of the following is NOT a true statement?

Ⓐ $-\frac{8}{9} > -\frac{9}{8}$, because $-\frac{8}{9}$ is located to the right of $-\frac{9}{8}$ on a number line.

Ⓑ $-3.5 < -2\frac{1}{4}$, because -3.5 is located to the left of $-2\frac{1}{4}$ on a number line.

Ⓒ $5\frac{1}{2} > -6.4$, because $5\frac{1}{2}$ is positive and -6.4 is negative.

Ⓓ $-4\frac{7}{8} > -3.5$, because $4\frac{7}{8} > 3.5$.

Unit 3 Assessment

Name _____

Standard 6.2(C) – Supporting

1 Between which pair of fractions is $\frac{5}{8}$ on the number line?

Ⓐ $\frac{5}{32}$ and $\frac{5}{16}$

Ⓑ $\frac{20}{32}$ and $\frac{10}{16}$

Ⓒ $\frac{17}{32}$ and $\frac{9}{16}$

Ⓓ $\frac{19}{32}$ and $\frac{11}{16}$

2 Which number line is correctly labeled?

Ⓕ 1 1.1 1.8 $1\frac{1}{2}$ $1\frac{2}{3}$ $1\frac{1}{4}$ 2

Ⓖ 1 $1\frac{1}{4}$ $1\frac{1}{2}$ $1\frac{2}{3}$ 1.1 1.8 2

Ⓗ 1 1.1 $1\frac{1}{4}$ $1\frac{2}{3}$ 1.8 $1\frac{1}{2}$ 2

Ⓙ 1 1.1 $1\frac{1}{4}$ $1\frac{1}{2}$ $1\frac{2}{3}$ 1.8 2

3 Which of the following best represents the point located on the number line below?

Ⓐ 1.5

Ⓑ −0.5

Ⓒ $-1\frac{1}{2}$

Ⓓ $\frac{1}{2}$

4 The average weight of a bag of gumballs from Charlie's Candy Factory is 8 ounces. The following table shows the difference between the average weight and the actual weight of each of five bags of gumballs.

Difference in Weight of Packages of Gumballs

Bag	Difference in Weight
1	0.1 oz
2	−0.13 oz
3	−0.08 oz
4	0.09 oz
5	−0.03 oz

Use the number line below to order the differences from greatest to least.

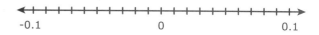

Ⓕ Bag 2, Bag 4, Bag 3, Bag 5, Bag 1

Ⓖ Bag 1, Bag 4, Bag 5, Bag 3, Bag 2

Ⓗ Bag 1, Bag 4, Bag 2, Bag 3, Bag 5

Ⓙ Bag 2, Bag 1, Bag 4, Bag 3, Bag 5

5 Which of the following is NOT a true statement?

Ⓐ $-0.5 < -\frac{1}{4}$, because −0.5 is located to the left of $-\frac{1}{4}$ on a number line.

Ⓑ $0.01 > -0.1$, because 0.01 is positive and −0.1 is negative.

Ⓒ $-5.96 > -\frac{1}{7}$, because $5.96 > \frac{1}{7}$.

Ⓓ $-\frac{5}{6} > -\frac{6}{5}$, because $-\frac{5}{6}$ is located to the right of $-\frac{6}{5}$ on the number line.

Name _____

Standard 6.2(C) – Supporting

Unit 3 Critical Thinking

1. Mrs. Byers challenges her students to write a number that is close to zero on a sheet of paper. Each group of four students then selects the student with the number closest to zero to stand in front of the class. The person with the number closest to zero will win a prize for their group. Four students stand in front of the class with the following numbers:

   Carlie: $\frac{1}{6}$, Billy: -0.14, Sam: 0.1, Allison: -0.2

Which student should win the prize for their group? Justify your answer. Include a number line or other model in your justification.

_____

_____

2. Consider the rational numbers $-\frac{5}{6}$ and $-\frac{2}{3}$. Which number is greater? _____ Does your answer change if the numbers are $\frac{5}{6}$ and $\frac{2}{3}$? Explain why or why not.

_____

_____

Is it possible to use a fraction model alone when comparing negative rational numbers? Why or why not?

_____

_____

_____

Unit 3 Journal/Vocabulary Activity

Name _____

Standard 6.2(C) – Supporting

Journal

Explain how using a number line can help in ordering rational numbers from least to greatest or greatest to least.

_____
_____
_____

Vocabulary Activity

Choose a word from the box below. Use the word to complete the graphic organizer.

Number line	Equal to	Positive
Greater than	Compare	Negative
Less than	Order	Inequality

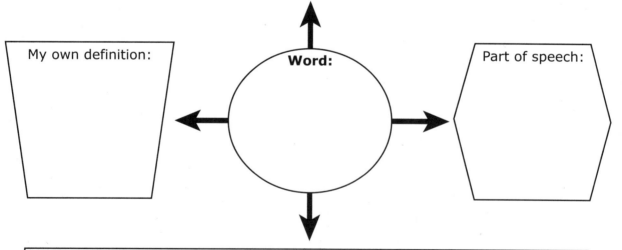

My own sentence:

My own definition:

Word:

Part of speech:

A picture that will remind me of what this word means to me:

28   ILLEGAL TO COPY   motivation**math**™ LEVEL 6   ©2014 mentoring**minds**.com

Standard 6.2(C) – Supporting

Unit 3 Motivation Station

Name That Phrase

Locate the point on the number line that corresponds to the given number. Label the point with the given letter to form a math phrase.

	Number	Letter
1.	$-\dfrac{1}{2}$	A
2.	−1.125	A
3.	0.375	B
4.	$\dfrac{1}{2}$	E
5.	$-\dfrac{7}{8}$	I
6.	$-\dfrac{3}{8}$	L
7.	0.25	M
8.	0	N
9.	$-\dfrac{5}{8}$	N
10.	−0.75	O
11.	$\dfrac{5}{8}$	R
12.	$-1\dfrac{1}{4}$	R
13.	$\dfrac{3}{4}$	S
14.	−1	T
15.	0.125	U

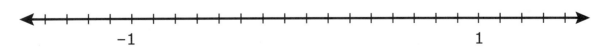

What phrase is formed by the points? _____

Unit 3 Homework

Name _____

Standard 6.2(C) – Supporting

1. Mr. Reyna's class orders rational numbers on a number line. Ricky draws cards with four rational numbers from a bag:

   $0.6, -4.25, 1\frac{3}{4}, -2\frac{3}{10}$

   Place Ricky's numbers on the number line below. Label each number above the line.

   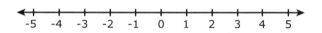

2. Using the numbers placed on the number line above, write three comparison statements using > or <.

   _____
   _____
   _____

3. Name an integer that is located to the left of ⁻4.25 on a number line.

   _____

   Write this number in the blank and place > or < in the circle to compare the numbers.

    ⁻4.25

4. Students in Mrs. Shupe's science class recorded high temperatures each day for a week in February. The table shows the data.

   Daily High Temperatures

Day	Temperature
Monday	2°F
Tuesday	11°F
Wednesday	−1°F
Thursday	0°F
Friday	−4°F

   Label the thermometer with the daily temperatures.

   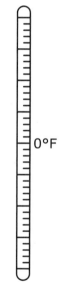

   List the temperatures in order from least to greatest.

   _____

   Which day of the week was the coldest?

   _____

   Which day of the week was the warmest?

   _____

Connections

Use a strip of butcher paper to create a number line from ⁻13 to +13. Using a standard deck of cards, assign positive values to black cards and negative values to red cards. Face cards are given values of 11, 12, and 13 for the Jack, Queen, and King cards, respectively. Deal five cards to each person and have them place the cards on the number line. Each player makes a comparison statement about two of his or her cards. For example, "⁻4 is greater than ⁻13." Continue ordering cards and making comparison statements until the entire deck is used.

Name _____

Standard 6.2(D) – Readiness

Unit 4 Introduction

1 Every year from 2010 to 2013, the low temperature in Juneau, Alaska, was recorded for April 1.

Temperature in Juneau on April 1

Year	Temperature
2010	-1.29°C
2011	-1.11°C
2012	-1.2°C
2013	-1.19°C

List the temperatures in order from warmest to coldest.

_____

2 Rick has homework in four subjects. The table shows the fraction of homework Rick has already completed for each subject.

Rick's Homework

Subject	Fraction Complete
Math	$\frac{1}{3}$
Science	$\frac{3}{5}$
Social Studies	$\frac{2}{3}$
Art	$\frac{5}{9}$

Rick plans to begin with the assignment that is the least completed and work his way to the most completed. In what order will Rick complete his homework assignments?

_____

_____

3 Record the set of numbers below in descending order.

$\frac{1}{4}$, 0.08, $\frac{3}{7}$, 82%

_____

4 Lorenzo has a collection of wrestling figures. The table shows the heights, in centimeters, of his favorite figures.

Heights of Wrestling Figures

Figure	Height (cm)
Hulk Higgins	11.1
Mr. Love	12
Dr. Macho	11.02
Fighting Fred	11.45

Order the heights of the figures from shortest to tallest.

_____

5 To make a purse, Sara needs four different prints of fabric, as shown in the table below.

Purse Fabrics

Print	Amount of Fabric Needed
Striped	$\frac{2}{5}$ yd
Polka dot	0.25 yd
Plaid	$\frac{3}{8}$ yd
Paisley	0.2 yd

List the fabric prints in order from greatest amount of fabric needed to least amount of fabric needed.

_____

6 List the rational numbers in ascending order.

$\frac{4}{5}$, -1.4, 0.75, -2$\frac{1}{3}$, 21%

_____

Unit 4 Guided Practice

Standard 6.2(D) – Readiness

1. Which list shows the numbers in descending order?

   Ⓐ $\frac{2}{3}$, 65%, $\frac{5}{8}$, 0.6

   Ⓑ $\frac{2}{3}$, $\frac{5}{8}$, 0.6, 65%

   Ⓒ 65%, $\frac{2}{3}$, 0.6, $\frac{5}{8}$

   Ⓓ $\frac{2}{3}$, 65%, 0.6, $\frac{5}{8}$

2. Mr. Campbell's shop class cut pieces of wood for a project. The table shows the lengths of wood cut by students.

Student	Length of Wood (inches)
Jeremy	$3\frac{1}{4}$
Shayla	$2\frac{7}{8}$
Melanie	$3\frac{5}{6}$
Ryan	$4\frac{2}{3}$
Chad	$3\frac{1}{2}$

   Which list shows the lengths of wood in order from shortest to longest?

   Ⓕ $2\frac{7}{8}$ in., $3\frac{1}{2}$ in., $3\frac{5}{6}$ in., $3\frac{1}{4}$ in., $4\frac{2}{3}$ in.

   Ⓖ $3\frac{1}{2}$ in., $2\frac{7}{8}$ in., $4\frac{2}{3}$ in., $3\frac{5}{6}$ in., $3\frac{1}{4}$ in.

   Ⓗ $2\frac{7}{8}$ in., $3\frac{1}{4}$ in., $3\frac{1}{2}$ in., $3\frac{5}{6}$ in., $4\frac{2}{3}$ in.

   Ⓙ $4\frac{2}{3}$ in., $3\frac{5}{6}$ in., $3\frac{1}{2}$ in., $3\frac{1}{4}$ in., $2\frac{7}{8}$ in.

3. Which list of integers is in order from greatest to least?

   Ⓐ 24, 9, -33, -24, -11

   Ⓑ -11, -18, -33, 9, 24

   Ⓒ 24, 9, -11, -24, -33

   Ⓓ -33, 24, -18, -11, 9

4. During a track meet, the following times, in seconds, were recorded for a 100-meter race. Which list shows the times in order from fastest to slowest?

   Ⓕ 15.3, 14.2, 14.0, 13.9, 14.8

   Ⓖ 13.9, 14.0, 14.2, 14.8, 15.3

   Ⓗ 15.3, 14.8, 14.2, 14.0, 13.9

   Ⓙ 13.9, 14.8, 14.2, 15.3, 14.0

5. The table shows the portion of sixth graders who passed their benchmark exams.

   Benchmark Exam Scores

Subject	Portion Passing Exam
English	82%
History	$\frac{4}{5}$
Math	$\frac{17}{20}$
Science	86%
Reading	$\frac{21}{25}$

   Which of the following lists the subjects in order from the least to greatest portion of students passing the benchmark exam?

   Ⓐ Science, Math, Reading, English, History

   Ⓑ English, History, Math, Reading, Science

   Ⓒ Math, History, Reading, English, Science

   Ⓓ History, English, Reading, Math, Science

Name _____

Standard 6.2(D) – Readiness

Unit 4 Independent Practice

1. The table shows the part of students in each of four classes who plan to attend the football game.

Football Game Attendance

Class	Part Attending the Football Game
Mrs. Quill's English	0.6
Mr. Tempo's Music	71%
Mr. King's History	$\frac{5}{6}$
Ms. Pascal's Math	56%

Which list shows the results in order from greatest to least?

Ⓐ 56%, 0.6, 71%, $\frac{5}{6}$

Ⓑ $\frac{5}{6}$, 71%, 0.6, 56%

Ⓒ $\frac{5}{6}$, 56%, 0.6, 71%

Ⓓ 56%, $\frac{5}{6}$, 0.6, 71%

2. Demarcus measured the lengths of four cords. Which list shows the lengths of the cords in order from shortest to longest?

Ⓕ $36\frac{1}{2}$ in., $32\frac{3}{4}$ in., $36\frac{7}{8}$ in., $32\frac{15}{16}$ in.

Ⓖ $32\frac{15}{16}$ in., $32\frac{3}{4}$ in., $36\frac{7}{8}$ in., $36\frac{1}{2}$ in.

Ⓗ $32\frac{3}{4}$ in., $32\frac{15}{16}$ in., $36\frac{1}{2}$ in., $36\frac{7}{8}$ in.

Ⓙ $36\frac{7}{8}$ in., $36\frac{1}{2}$ in., $32\frac{15}{16}$ in., $32\frac{3}{4}$ in.

3. Which lists the numbers in ascending order?

Ⓐ $\frac{1}{3}$, 0.3, 25%, $\frac{2}{5}$

Ⓑ 25%, 0.3, $\frac{1}{3}$, $\frac{2}{5}$

Ⓒ $\frac{1}{3}$, $\frac{2}{5}$, 0.3, 25%

Ⓓ 0.3, $\frac{1}{3}$, $\frac{2}{5}$, 25%

4. On January 1, Rodney recorded the outside temperature at four different times. The temperatures are shown in the table below.

Outside Temperatures

Time	Temperature (°F)
6:00 A.M.	-8
11:00 A.M.	10
4:00 P.M.	0
9:00 P.M.	-2

Which list shows the temperatures in order from coldest to warmest?

Ⓕ 0°F, -2°F, -8°F, 10°F

Ⓖ -2°F, -8°F, 0°F, 10°F

Ⓗ -8°F, -2°F, 0°F, 10°F

Ⓙ -8°F, 0°F, -2°F, 10°F

5. Tyler is competing in the discus event at the city track meet. The table below lists his daily best throws.

Tyler's Discus Throws

Day	Length (in meters)
Monday	167.468
Tuesday	167.079
Wednesday	167.5
Thursday	167.48

Which list shows the days in order from the longest throw to the shortest throw?

Ⓐ Monday, Tuesday, Thursday, Wednesday

Ⓑ Wednesday, Thursday, Tuesday, Monday

Ⓒ Tuesday, Monday, Thursday, Wednesday

Ⓓ Wednesday, Thursday, Monday, Tuesday

Unit 4 Assessment

Standard 6.2(D) – Readiness

1. Marquis, Chad, Jose, and Stephen competed to see who could run the farthest without stopping. Marquis ran $\frac{1}{2}$ mile, Chad ran $\frac{7}{8}$ mile, Jose ran $\frac{3}{5}$ mile, and Stephen ran $\frac{3}{4}$ mile. Which list shows these distances in order from greatest to least?

   Ⓐ $\frac{7}{8}$ mi, $\frac{3}{5}$ mi, $\frac{1}{2}$ mi, $\frac{3}{4}$ mi

   Ⓑ $\frac{7}{8}$ mi, $\frac{3}{4}$ mi, $\frac{3}{5}$ mi, $\frac{1}{2}$ mi

   Ⓒ $\frac{3}{4}$ mi, $\frac{1}{2}$ mi, $\frac{3}{5}$ mi, $\frac{7}{8}$ mi

   Ⓓ $\frac{1}{2}$ mi, $\frac{3}{4}$ mi, $\frac{7}{8}$ mi, $\frac{3}{5}$ mi

2. Four stores are having sales. The table shows the price reduction at each store.

   Store Sales

Store Name	Price Reduction
Zell's	$\frac{1}{3}$
Mallard's	30%
Tracey's	25%
Jenny's	$\frac{2}{5}$

   Which list shows the price reductions from least savings to greatest savings?

   Ⓕ 25%, 30%, $\frac{1}{3}$, $\frac{2}{5}$

   Ⓖ $\frac{1}{3}$, 25%, 30%, $\frac{2}{5}$

   Ⓗ 25%, $\frac{1}{3}$, 30%, $\frac{2}{5}$

   Ⓙ $\frac{2}{5}$, 25%, 30%, $\frac{1}{3}$

3. Which list of integers is in descending order?

   Ⓐ 3, 0, -14, -8, -5, -3

   Ⓑ 0, 3, -3, -5, -8, -14

   Ⓒ -14, -8, -5, -3, 3, 0

   Ⓓ 3, 0, -3, -5, -8, -14

4. Maritza cut strips of ribbon and measured their lengths in centimeters. Which list shows the lengths in order from longest to shortest?

   Ⓕ 26.1 cm, 26 cm, 25.69 cm, 25.7 cm

   Ⓖ 25.69 cm, 25.7 cm, 26 cm, 26.1 cm

   Ⓗ 25.7 cm, 25.69 cm, 26 cm, 26.1 cm

   Ⓙ 26.1 cm, 26 cm, 25.7 cm, 25.69 cm

5. Braden took a survey of his classmates to determine which electives they are taking. The table below shows the results of his survey.

   Sixth-Grade Electives

Elective	Part of Class
Choir	0.35
Art	$\frac{4}{9}$
Band	42%
Theater	$\frac{3}{8}$

   Which list shows the electives in order from most popular to least popular based on the survey results?

   Ⓐ Band, Art, Theater, Choir

   Ⓑ Art, Band, Theater, Choir

   Ⓒ Band, Choir, Art, Theater

   Ⓓ Art, Band, Choir, Theater

Name _____

Standard 6.2(D) – Readiness

Unit 4 Critical Thinking

1 In ascending order, name an integer, fraction, decimal, and percent that are between 0.99 and 1.01.

0.99, _____ , _____ , _____ , _____ , 1.01

In ascending order, name an integer, fraction, decimal, and percent that are between -1.01 and -0.99.

-1.01, _____ , _____ , _____ , _____ , -0.99

Explain how you determined your answers.

_____
_____
_____
_____
_____
_____

2 Are the fractions below listed in order from least to greatest?

$$-\frac{1}{4}, -\frac{1}{5}, -\frac{1}{6}, -\frac{1}{7}, -\frac{1}{8}$$

Answer: _____

Justify your answer using a number line.

Unit 4 Journal/Vocabulary Activity

Standard 6.2(D) – Readiness

Journal

Explain how to order a set of numbers that includes at least one fraction, one percent, and one decimal number.

_____

_____

_____

Vocabulary Activity

Read each statement below. Decide if the numbers in each situation would be ordered from smallest to largest or largest to smallest. Place the number of each statement in the correct box.

Order the Numbers

Smallest to Largest	Largest to Smallest

1. Mrs. Marshall's students are running a race. Order their times from fastest to slowest.

2. Kara measures the heights of five plants. Order the heights from shortest to tallest.

3. Mr. Todd recorded the temperature each day last week. Arrange the temperatures from warmest to coldest.

4. Ms. Clark posts students' grades weekly. Record the students' grades in ascending order.

5. Mrs. Teasley weighs each bag of fruit before placing it in her shopping cart. List the weights of the bags from heaviest to lightest.

6. Erin and her three sisters ran as far as they could without stopping. Arrange the distances the girls ran from shortest to farthest.

7. Mike tracks the amount of money he spends each day. List his daily spending in order from greatest to least.

8. Harris and four of his friends go fishing. Record the lengths of the fish caught in descending order.

Name _____

Standard 6.2(D) – Readiness

Unit 4 Motivation Station

Connect the Dots

Start at $-\frac{9}{2}$. Use a pencil to connect the dots of the rational numbers in order from least to greatest.

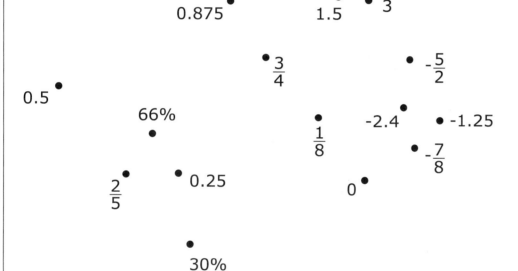

What picture do you see?

Unit 4 Homework

Standard 6.2(D) – Readiness

1 The following table shows the times for athletes running a 100-meter race.

100-Meter Race

Runner	Time (seconds)
Mary	0.34
Rhea	0.394
Lindsay	0.402
Mel	0.334
Brooke	0.35

If the first three places receive ribbons, which runners received ribbons for the race?

1st Place Ribbon: _____

2nd Place Ribbon: _____

3rd Place Ribbon: _____

2 The science club sold cold drinks after school to earn money for the science fair. They sold $3\frac{1}{2}$ gallons of soda, $2\frac{3}{4}$ gallons of orange juice, $3\frac{3}{8}$ gallons of lemonade, and $3\frac{1}{3}$ gallons of limeade. List the drinks in order from greatest to least amount sold.

_____

_____

3 Maria measured the following ribbon lengths.

Ribbon Lengths

Color of Ribbon	Length (feet)
Red	$\frac{7}{12}$
Green	0.75
Purple	$\frac{5}{6}$
Blue	0.375

Arrange the ribbon lengths in order from shortest to longest.

_____

4 The Highland Mall manager created a report showing the percent change in sales at five stores.

Sales Report

Store	Percent Change
Navy Blues	−1.6%
Soaring Eagles	+1.66%
Unlimited Too	−1.06%
Tot Togs	−1.606%
Bear Facts	+1.066%

List the percent change in sales from least to greatest for all five stores.

_____

_____

Connections

1. Use a standard deck of cards and remove all the face cards. All red cards represent negative numbers and all black cards represent positive numbers. Turn over five cards at a time and arrange them from least to greatest or greatest to least.

2. Use a newspaper or magazine to find five rational numbers. Cut them out and then arrange the rational numbers from least to greatest or greatest to least.

Standard 6.2(E) – Supporting

Unit 5 Introduction

1 Eight members of the math club equally shared three pizzas. Expressed as a fraction, what part of a pizza did each member receive?

_____

Write the part of a pizza each member received as a decimal.

_____

Draw a picture to represent your answer.

2 During a science investigation, fifteen students divided 6 pounds of potting soil equally. Write an expression that shows how many pounds of potting soil each student received.

_____

Written as a decimal, how many pounds of potting soil did each student receive?

_____

3 Chin brought a dozen cupcakes, and Ming brought 36 cookies to school for the Western Day celebration. The cupcakes and cookies are each divided equally among the 24 students in the class. What part of the cupcakes does each student receive?

_____

4 Margaret uses 80 yards of fabric to make costumes for the school play. She needs to make 25 costumes with the fabric and will divide the fabric equally among the costumes. Write an equation to represent the number of yards of fabric, f, for each costume.

_____

Expressed as a decimal, how many yards of fabric does Margaret use for each costume?

_____

5 Maddie correctly answered 17 out of 20 questions on her math test. Expressed as a decimal, what part of the questions did Maddie answer correctly?

_____

6 Logan split a rope that was 28 inches long into 5 equal parts. Leroy split a rope that was 32 inches long into 6 equal parts. Which boy's rope was cut into longer pieces?

_____

Explain your answer.

_____
_____
_____

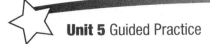

Unit 5 Guided Practice

Standard 6.2(E) – Supporting

1. At Becky's Restaurant, 10 out of 25 customers order wheat rolls. The rest of the customers order cornbread. Which does NOT represent the part of the customers who order wheat rolls?

   A $\frac{10}{25}$

   B $\frac{2}{5}$

   C $25 \div 10$

   D 0.4

2. Jane and Joe's mom cooks sausage links for breakfast. There are 7 links to be shared equally between the children. Which equation shows how Jane and Joe can determine the number of sausage links they each receive?

   F $2 \div 7 = 0.29$

   G $2 \div 7 = 3.5$

   H $7 \div 2 = 3.1$

   J $7 \div 2 = 3.5$

3. Armia says he can determine the quotient of $\frac{121}{11}$ by solving the expression $11 \div 121$. Is Armia correct?

   A Yes, because $\frac{121}{11} = 11$ and $11 \div 121 = 11$

   B Yes, because $\frac{121}{11} = \frac{1}{11}$ and $11 \div 121 = \frac{1}{11}$

   C No, because $\frac{121}{11} = 11$ and $11 \div 121 = \frac{1}{11}$

   D No, because $\frac{121}{11} = \frac{1}{11}$ and $11 \div 121 = 11$

4. Of the 350 seniors at Westside High School, $\frac{1}{25}$ plan to attend junior college in the fall. Which represents the part of the seniors who plan to attend junior college?

   F $1 \overline{) 25}$

   G 0.40

   H $\frac{25}{350}$

   J 0.04

5. Mrs. Hall makes sack lunches for the 9 children in her scout group. She has 3 bags of chips to divide equally among the lunches. Which shows the part of a bag of chips that should go in each lunch?

   A $\frac{9}{3}$

   B $0.\overline{3}$

   C 3

   D $\frac{2}{3}$

6. Jim has a box of 9 granola bars to share among the 4 children in his family. He wants to make sure that each child receives exactly the same number of granola bars. Jim first gives each person 2 granola bars. What should Jim do next?

   F Divide 1 bar into 4 equal pieces, so that each person receives a total of $2\frac{1}{4}$ granola bars.

   G Divide 1 bar into 2 equal pieces, so that each person receives a total of $2\frac{1}{2}$ granola bars.

   H Divide 9 bars into 2 equal pieces, so that each person receives a total of $4\frac{1}{2}$ granola bars.

   J Divide 9 bars into 8 equal pieces, so that each person receives a total of $1\frac{1}{8}$ granola bars.

Standard 6.2(E) – Supporting

Unit 5 Independent Practice

1 Amber calculates that $\frac{1}{3}$ of the students in her fourth-period math class are wearing flip-flops. Which best represents the part of her class that is wearing flip-flops?

Ⓐ $3 \div 1$

Ⓑ 0.03

Ⓒ $0.\overline{3}$

Ⓓ $1\overline{)3}$

2 Alejandro has 8 cans of soda that he wants to share equally with 16 friends. Which expressions can be used to determine the amount of soda each friend will receive?

  I. $\frac{16}{8}$

  II. $8\overline{)16}$

  III. $8 \div 16$

  IV. $\frac{8}{16}$

Ⓕ I and II only

Ⓖ III and IV only

Ⓗ II, III, and IV only

Ⓙ I, II, III, and IV

3 Joe records statistics for his favorite baseball player.

Statistics

Average	Hits	On Base	Slugging
0.270	0.500	0.350	0.600

Of the statistics listed in the table, which represents the quotient of $\frac{7}{20}$?

Ⓐ Average

Ⓑ Hits

Ⓒ On Base

Ⓓ Slugging

4 A 50-minute block of computer time must be shared equally among Maleeka and two other students in Mrs. Moore's class. Which equation best represents Maleeka's computer time in minutes?

Ⓕ $\frac{3}{50} = 16.\overline{6}$

Ⓖ $\frac{50}{3} = 16.2$

Ⓗ $\frac{50}{3} = 16.04$

Ⓙ $\frac{50}{3} = 16.\overline{6}$

5 Montrel plays the saxophone. He plans to practice $\frac{3}{2}$ as many days as last year. Which does NOT represent the change in Montrel's practice time?

Ⓐ $2\overline{)3}$

Ⓑ 1.2

Ⓒ $1\frac{1}{2}$

Ⓓ $3 \div 2$

6 The six members of the Arnold family calculate the family's daily water use to be 38 gallons. Which equation best shows the average amount of water each person uses daily?

Ⓕ $6 \div 38 = \frac{3}{19}$

Ⓖ $38 \div 6 = 6\frac{1}{3}$

Ⓗ $\frac{6}{38} = 0.15$

Ⓙ $\frac{38}{6} = 6.03$

Unit 5 Assessment

Standard 6.2(E) – Supporting

1. Five brothers equally shared 3 boxes of snack crackers. Which equation represents the part of the boxes that each brother received?

   A) $\frac{5}{3} = 1\frac{2}{3}$

   B) $\frac{5}{3} = 1.6$

   C) $\frac{3}{5} = 0.6$

   D) $\frac{3}{5} = 0.\overline{6}$

2. Yesterday, 15 out of 20 students attended choir practice. Which does NOT represent the part of the students who attended choir practice?

   F) 0.75

   G) $15 \div 20$

   H) $15\overline{)20}$

   J) $\frac{3}{4}$

3. Nevaeh shares 8 candy bars equally among her 12 friends. Which expressions can be used to determine the part of the candy bars each friend receives?

   I. $\frac{8}{12}$

   II. $8\overline{)12}$

   III. $8 \div 12$

   IV. $0.\overline{6}$

   A) I and II only

   B) III and IV only

   C) I, III, and IV only

   D) I, II, III, and IV

4. Ramon correctly answered 0.84 of the questions on his test. Which quotient represents the part of Ramon's test questions that were answered correctly?

   F) $\frac{21}{25}$

   G) $\frac{84}{10}$

   H) $\frac{17}{20}$

   J) $\frac{7}{8}$

5. In Myron's class of 24 students, 9 students plan to travel during spring break. Which represents the part of the students that plan to travel during spring break?

   A) $24 \div 9$

   B) $\frac{1}{3}$

   C) 0.30

   D) 0.375

6. A group of 6 fifth graders read a total of 29 books. A group of 8 sixth graders read a total of 33 books. Which group won the prize for reading the most books per person?

   F) The fifth graders won, because $4\frac{5}{6} > 4\frac{1}{8}$.

   G) The sixth graders won, because $4\frac{1}{8} > 4\frac{5}{6}$.

   H) The sixth graders won, because $33 > 29$.

   J) The teams tied, because they each read $4\frac{1}{2}$ books per person.

Standard 6.2(E) – Supporting

Unit 5 Critical Thinking

1. Follow the directions for each word problem.

   Mrs. Tanner has a piece of yarn that is 4 inches long. She wants to cut the yarn into 3 equal segments for a craft project. Write an expression three different ways to determine the length, in inches, of each segment.

   _____

   Record the length of each segment as a decimal.

   _____

   Mr. Tanner needs bags of jelly beans to give to each of his 4 sons. He was only able to find one 3-pound package of jelly beans. Write an expression three different ways to determine the amount of jelly beans, in pounds, that should go in each bag.

   _____

   Record the weight of jelly beans that will go in each bag as a decimal.

   _____

   Explain how the expressions used to solve the word problems are similar and how they are different.

   _____
   _____
   _____

2. Using whole numbers, write a division word problem that has a quotient of 0.125.

   _____
   _____
   _____
   _____

Unit 5 Journal/Vocabulary Activity

Name _____

Standard 6.2(E) – Supporting

Journal

Explain how fractions are related to division.

_____

_____

Verify your explanation using an example or illustration.

Vocabulary Activity

ACROSS

1. A quantity to be divided
5. The expression written above the fraction bar to indicate the number of equal parts being described
6. The operation of determining how many times one quantity is contained in another; the inverse of multiplication
7. An expression that indicates the quotient of two quantities

DOWN

2. The expression written below the fraction bar that indicates the number of equal parts into which one whole is divided
3. The answer to a division problem
4. The quantity by which another quantity, the dividend, is to be divided

Name _____

Standard 6.2(E) – Supporting

Unit 5 Motivation Station

Claim and Color

Play *Claim and Color* with a partner. Each pair of players needs a game board, a pencil, and a paper clip to use with the spinner. Each player needs a different color of marker or colored pencil to shade squares on the game board. Player 1 spins the spinner and selects a square on the game board that is equivalent to the number spun. If the square is selected correctly, player 1 colors the square to claim it. If incorrect, player 1 does not color a square for the turn. Play passes to player 2 who repeats the process. If there are no more squares that are equal to the number spun, the player loses a turn. Play continues until the teacher calls time. The player with more colored squares is the winner.

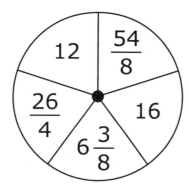

$\dfrac{96}{6}$	$\dfrac{60}{5}$	$\dfrac{27}{4}$	$\dfrac{48}{4}$	6.75
$\dfrac{36}{3}$	$\dfrac{13}{2}$	$6\dfrac{1}{2}$	$\dfrac{80}{5}$	$\dfrac{102}{16}$
$6\dfrac{6}{8}$	$\dfrac{51}{8}$	6.125	$\dfrac{64}{4}$	$\dfrac{84}{7}$
6.5	$\dfrac{48}{3}$	$\dfrac{96}{8}$	$6\dfrac{3}{4}$	6.05
$\dfrac{32}{2}$	$\dfrac{72}{6}$	6.375	$\dfrac{52}{8}$	$6\dfrac{2}{4}$

©2014 mentoringminds.com   motivationmath™ LEVEL 6   ILLEGAL TO COPY

Unit 5 Homework

Name _____

Standard 6.2(E) – Supporting

1. One dozen teachers shared 7 ounces of chocolate equally. How many ounces of chocolate did each teacher receive? Record your answer as a fraction and a decimal.

   _____

2. Mrs. Miller asks her students to simplify the expression $\frac{10}{16}$. Jack records an answer of $\frac{5}{8}$. His neighbor, Landon, writes 0.625. Who is correct?

   _____

   Explain.

   _____
   _____

3. A dance teacher makes bows for her students' costumes. She has 24 yards of ribbon and cuts the ribbon into 36 equal pieces. Expressed as a fraction, how many yards long is each piece of ribbon?

   _____

   What is the length, in yards, expressed as a decimal?

   _____

4. Rosanne purchases 3 dozen boxes of paper clips for her office. She divides the paper clips equally among 5 containers. Write an equation to represent the number of boxes, *n*, of paper clips in each container.

   _____

   Expressed as a decimal, how many boxes of paper clips are in each container?

   _____

5. Micah and 3 friends went fishing on Saturday. Micah caught 9 fish, Sara caught 4 fish, Diego caught 6 fish, and Emily caught 7 fish. Since all the fish were the same size, the group decided to share them equally. Expressed as a decimal, how many fish did each person take home?

   _____

6. At Dr. Carter's dental office, 6 out of every 10 patients in the waiting room are there for a check-up and cleaning. Represent the part of Dr. Carter's patients who are waiting for a check-up and cleaning as a quotient of two numbers and as a decimal.

   _____

Connections

1. Using three sticks of gum, determine how many sticks of gum, or parts of a stick, each person in your family would receive if the gum is shared equally.

2. Think of real-world situations when it is best to leave the quotient of two numbers as a fraction and when it is best to express the quotient as a decimal.

Standard 6.3(A) – Supporting

Unit 6 Introduction

1. Write the reciprocal for each of the following.

    $\frac{5}{8}$ _____ $3\frac{1}{2}$ _____

    $\frac{1}{4}$ _____ 15 _____

    $\frac{5}{6}$ _____ $\frac{9}{8}$ _____

2. Fatima has 42 coins to separate into 7 equal groups.

    Write a numeric equation using division to show the number of coins in each group.

    _____

    Write a numeric equation using multiplication to show the number of coins in each group.

    _____

3. Jessie and Vicente have a bag of treats to share equally among a group of puppies. There are 54 treats in the bag, and they have 6 puppies. Jessie wants to determine the number of treats each puppy should receive by dividing. Vicente wants to use multiplication to find the number of treats each puppy should get. Who is correct?

    Justify your answer and show the equation or equations you use.

    _____
    _____
    _____

4. Write four multiplication equations to determine the number of evenly distributed groups that can be created from a bundle of 24 balloons.

    _____
    _____
    _____
    _____

5. A developer purchases a 45-acre plot of land. He wants to subdivide the land into $\frac{3}{4}$-acre lots to build homes. Write two expressions the developer can use to determine the number of houses that can be built in the subdivision.

    _____
    _____

    How many houses can be built?

    _____

6. Marissa has collected 200 colored beads to make bracelets for her friends. Each bracelet requires 24 colored beads. Marissa writes the following expression to determine the number of bracelets she can make:

    $$200 \div 24$$

    What is another expression Marissa could use to find the number of bracelets she can make?

    _____

    How many bracelets can Marissa make?

    _____

Unit 6 Guided Practice

Standard 6.3(A) – Supporting

1. Which of the following expressions has the same solution?

   I. $72 \div 9$

   II. $9 \div 72$

   III. $72 \cdot \frac{1}{9}$

   IV. $72 \div \frac{1}{9}$

   Ⓐ I and II only

   Ⓑ I and IV only

   Ⓒ I and III only

   Ⓓ II, III, and IV only

2. Kenedi has a piece of ribbon that measures $24\frac{1}{4}$ inches long. She cuts the ribbon into 15 equal pieces to attach to her dance costume. Which of the following expressions CANNOT be used to determine the length of each piece of ribbon Kenedi uses on her costume?

   Ⓕ $\frac{97}{4} \div \frac{1}{15}$

   Ⓖ $\left(24\frac{1}{4}\right)\left(\frac{1}{15}\right)$

   Ⓗ $\frac{97}{4} \div 15$

   Ⓙ $24\frac{1}{4} \div 15$

3. Nancy has 5 pounds of jelly beans to divide into bags for her grandchildren. If she puts $\frac{1}{4}$ pound of jelly beans in each bag, how many bags can she make?

   Ⓐ 10 bags

   Ⓑ 5 bags

   Ⓒ $1\frac{1}{4}$ bags

   Ⓓ 20 bags

4. Which of the models below shows the same solution as $\frac{2}{3} \div \frac{1}{2}$?

   Ⓕ 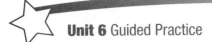 $= \frac{1}{3}$

   Ⓖ  $= 1\frac{1}{3}$

   Ⓗ 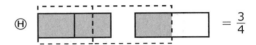 $= \frac{3}{4}$

   Ⓙ 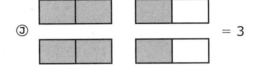 $= 3$

5. After the Diamond Divas' last softball game, the team went to John's Pizza Palace for a party. They ordered 8 pizzas to be shared among the team. If each player received $\frac{1}{2}$ of a pizza, how many players attended the party?

   Record your answer and fill in the bubbles on the grid below. Be sure to use the correct place value.

Name _____

Standard 6.3(A) – Supporting

Unit 6 Independent Practice

1. Which of the following expressions is equivalent to $12 \div \frac{3}{2}$?

   I. $\frac{1}{12}\left(\frac{2}{3}\right)$

   II. $12\left(\frac{3}{2}\right)$

   III. $\frac{1}{12}\left(\frac{3}{2}\right)$

   IV. $12\left(\frac{2}{3}\right)$

   Ⓐ I and IV only

   Ⓑ I and III only

   Ⓒ II and IV only

   Ⓓ IV only

2. Which of the following shows the correct solution for the expression $\frac{5}{2} \div \frac{2}{3}$?

   Ⓕ $\frac{5}{2} \cdot \frac{2}{3} = \frac{5}{3}$

   Ⓖ $\frac{5}{2} \cdot \frac{3}{2} = \frac{15}{4}$

   Ⓗ $\frac{2}{5} \cdot \frac{2}{3} = \frac{4}{15}$

   Ⓙ $\frac{2}{5} \cdot \frac{3}{2} = \frac{3}{5}$

3. Mrs. Rucker and Mrs. Overall are given 66 calculators to share equally between their classrooms. Which of the following expressions can be used to determine the number of calculators each teacher receives?

   Ⓐ $66 \cdot \frac{1}{2}$, because the expression is equivalent to $66 \div 2$

   Ⓑ $66 \div \frac{1}{2}$, because the expression is equivalent to $66 \cdot \frac{1}{2}$

   Ⓒ $66 \cdot 2$, because the expression is equivalent to $66 \div \frac{1}{2}$

   Ⓓ $66 \div 2$, because the expression is equivalent to $66 \cdot 2$

4. There are 36 students in a class. The teacher wants to create groups with an equal number of students in each group. Which of the following expressions could NOT be used to determine the number of groups the teacher can create?

   Ⓕ $36 \cdot \frac{1}{4}$

   Ⓖ $36 \div 6$

   Ⓗ $36 \cdot \frac{1}{9}$

   Ⓙ $36 \div \frac{1}{12}$

5. Josiah calculates his average for social studies class. The sum of his grades on all of the assignments is 482. There are a total of 6 assignments in the class. To find his average, Josiah multiplies 482 by $\frac{1}{6}$. Which of the following would give Josiah the same result?

   Ⓐ Dividing 482 by $\frac{1}{6}$

   Ⓑ Multiplying $\frac{1}{482}$ by 6

   Ⓒ Dividing 482 by 6

   Ⓓ Multiplying $\frac{1}{482}$ by $\frac{1}{6}$

6. Liam's family traveled a distance of 504.5 miles at a speed of 65 miles per hour. Which of the following expressions could NOT be used to determine the number of hours it took Liam's family to travel 504.5 miles?

   Ⓕ $504.5 \div 65$

   Ⓖ $504.5\left(\frac{1}{65}\right)$

   Ⓗ $65\left(\frac{1}{504.5}\right)$

   Ⓙ $\frac{504.5}{65}$

Unit 6 Assessment

Name _____

Standard 6.3(A) – Supporting

1 Which of the following is NOT equivalent to the expression $15 \cdot \frac{4}{3}$?

I. $15 \div \frac{3}{4}$
II. $15 \cdot 1\frac{1}{3}$
III. $15 \div \frac{4}{3}$
IV. $15 \cdot \frac{3}{4}$

Ⓐ I and II only

Ⓑ II only

Ⓒ III and IV only

Ⓓ I and IV only

2 Louise has a 5-pound bag of rice she needs to separate into single servings. A serving is $\frac{1}{4}$ cup or $\frac{1}{8}$ pound. Which of the following shows how to find the number of servings the bag of rice contains?

Ⓕ $5\left(\frac{1}{8}\right) = \frac{5}{8}$

Ⓖ $5(8) = 40$

Ⓗ $5\left(\frac{1}{4}\right) = 1\frac{1}{4}$

Ⓙ $5(4) = 20$

3 An airplane flying at an altitude of 30,000 feet begins to descend at a rate of 1,500 feet per minute. Which expression can be used to determine how many minutes it will take the plane to reach the ground, if it continues to descend at the same rate?

Ⓐ $30{,}000 \div \frac{1}{1{,}500}$

Ⓑ $\frac{1}{30{,}000} \div 1{,}500$

Ⓒ $\frac{1}{30{,}000} \cdot 1{,}500$

Ⓓ $30{,}000\left(\frac{1}{1{,}500}\right)$

4 Which of the following statements is true?

Ⓕ Dividing by a rational number is equivalent to dividing the dividend by the reciprocal of the rational number.

Ⓖ Dividing by a rational number is equivalent to multiplying the dividend by the reciprocal of the rational number.

Ⓗ Dividing by a rational number is equivalent to dividing the reciprocal of the dividend by the rational number.

Ⓙ Dividing by a rational number is equivalent to multiplying the reciprocal of the dividend by the rational number.

5 Kay has $1\frac{1}{4}$ hours to complete 3 chores before leaving for soccer practice. If she spends an equal amount of time on each chore, which of the following could Kay use to find the part of an hour each chore will require?

Ⓐ $\frac{5}{4} \cdot \frac{1}{3}$

Ⓑ $\frac{5}{4} \cdot 3$

Ⓒ $\frac{4}{5} \cdot \frac{1}{3}$

Ⓓ $\frac{4}{5} \cdot 3$

6 Ryan has $4\frac{2}{3}$ yards of fabric to make clothes for dolls she sells at the trade show. Each pattern requires $1\frac{1}{6}$ yards of fabric. Which equation shows the number of dolls Ryan will be able to clothe for the trade show?

Ⓕ $\frac{3}{14} \div \frac{6}{7} = \frac{1}{4}$

Ⓖ $\frac{14}{3} \div \frac{6}{7} = 5\frac{4}{9}$

Ⓗ $\frac{3}{14} \div \frac{7}{6} = \frac{9}{49}$

Ⓙ $\frac{14}{3} \div \frac{7}{6} = 4$

Name _____

Standard 6.3(A) – Supporting

Unit 6 Critical Thinking

1. Draw a model for each of the problems shown. Then complete each equation by writing the correct answer in the blank.

$\frac{2}{4} \div \frac{1}{4} = $ _____ $\frac{2}{4} \cdot 4 = $ _____

Explain how these problems are alike and how they are different.

_____

_____

2. Use the following expression to write and solve a word problem. Show all work on solution.

$$\frac{5}{8} \div 2\frac{1}{2}$$

Name _____

Unit 6 Journal/Vocabulary Activity

Standard 6.3(A) – Supporting

Journal

How is dividing 4 by $\frac{1}{4}$ different from multiplying 4 by $\frac{1}{4}$?

_____
_____
_____
_____

Vocabulary Activity

For each box below, fill in the following information: *Definition* (math definition), *In your own words* (definition the way it makes sense to you), *Examples*, and *Non-examples*. Give at least two examples and non-examples for each word.

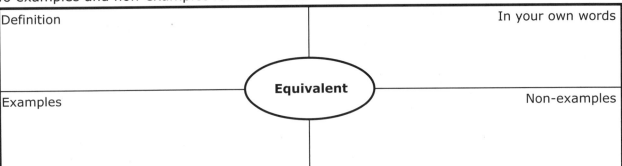

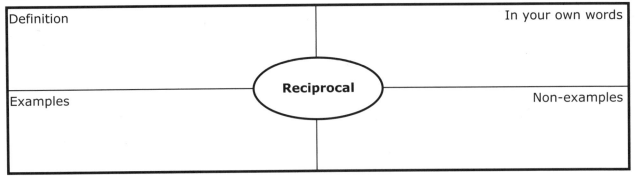

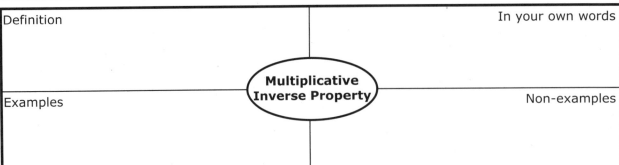

Expression Match

Play *Expression Match* with a partner. Each pair of players needs a game board, a number cube, and a paper clip to use with the spinner. Each player needs two centimeter cubes of the same color and a pencil. Player 1 rolls the number cube and spins the spinner below. Beginning at the START positions, player 1 moves one cube clockwise the number of spaces shown on the number cube. The second cube is moved counter clockwise on the inner circle the number of spaces shown on the spinner. If the two expressions are a match, player 1 writes his/her initials on both spaces and rolls/spins again. If the expressions do not match, play passes to player 2. For each turn, play proceeds from the last position of that player's cubes. Play continues until time is called. The winner is the player with more initialed spaces on the board.

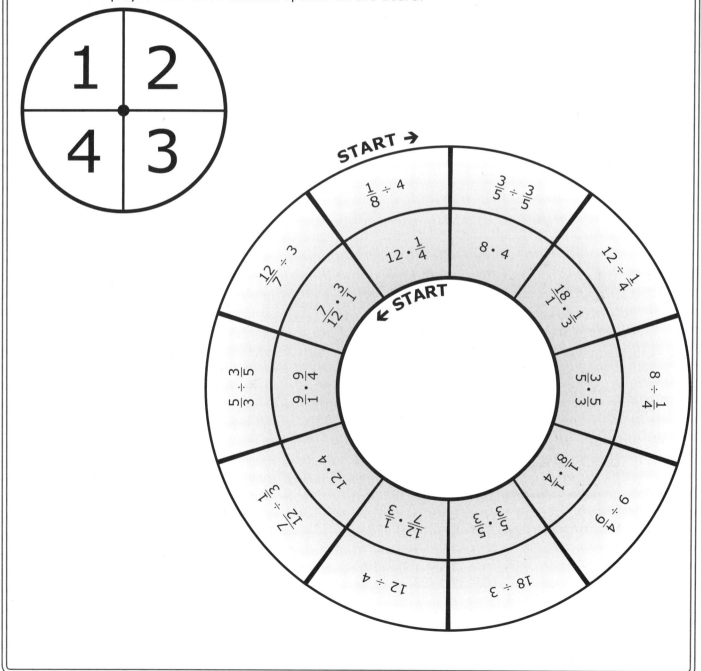

Unit 6 Homework

Standard 6.3(A) – Supporting

1. Janice has a string that is $2\frac{1}{4}$ feet long. She needs to cut the string into equal sections measuring $\frac{1}{8}$ foot each. Write and solve two different equations showing how to find the number of sections that can be cut from the string.

2. Write one multiplication equation and one division equation to represent the following situation.

    32 animal crackers shared by 8 friends

3. Dez and Angie bought a bag containing 300 peanuts. They want to share the peanuts equally. Dez divides 300 by 2. Angie multiplies 300 by $\frac{1}{2}$. Who performs the correct operation?

    Explain your answer.

4. Abraham has $2\frac{1}{2}$ pounds of food to share between his 5 cats. He believes he should multiply $2\frac{1}{2}$ by $\frac{1}{5}$ to find how much food each cat will receive. His friend Amelia disagrees and says he should multiply $\frac{2}{5}$ by 5 to find how much food each cat will receive. Who is correct?

    Write an equation and draw a model to support your answer.

5. The Johnson family drives around the country on their summer vacation. They traveled a total of 2,750 miles in $12\frac{1}{2}$ days. Write and solve two equations to show the average number of miles the Johnson family drives per day.

Connections

Practice writing reciprocals of numbers, and teach the members of your family. Find numbers in your home by looking at recipes, measurements, newspapers, or magazines. Practice with numbers that are fractions and whole numbers. Take turns quizzing each other once everyone has learned.

Standard 6.3(B) — Supporting

Unit 7 Introduction

1 Marty plans to make an alien costume for a party. He needs $\frac{7}{8}$ yard of fabric and $1\frac{1}{2}$ yards of aluminum foil for the costume. Fabric costs $5 per yard. Will the fabric needed to make the costume cost more than $5 or less than $5? _____

Explain your answer.

_____

_____

2 Yesterday, Ricardo ran one lap around the school track in 180 seconds. Today, he ran one lap in $\frac{6}{5}$ of yesterday's time. Did Ricardo's time increase or decrease from yesterday to today? _____

Explain how you found your answer.

_____

_____

3 Marilyn's garden is 3 yards long and $\frac{2}{3}$ yard wide. Carolyn's garden is 3 yards long and $\frac{3}{2}$ yards wide. Is the area of Marilyn's garden larger or smaller than 3 square yards?

_____

Explain your answer.

_____

_____

Is the area of Carolyn's garden larger or smaller than 3 square yards?

_____

Explain your answer.

_____

_____

4 Use >, <, or = to make the statement true.

$$5 \times \frac{8}{9} \_\_\_\_\_ 5$$

Justify your answer by completing the computation.

5 Greg has a blue sticky note and a pink sticky note. The blue sticky note has an area of x square inches. The pink sticky note has an area that is 0.75 times the area of the blue sticky note. Write a statement comparing the areas of the blue and pink sticky notes.

_____

_____

_____

6 Which of the following expressions results in a product greater than 45? Circle your answer(s).

    I. $\frac{4}{5} \times 45$

   II. $45 \cdot 4.5$

  III. $45 \times \frac{5}{4}$

  IV. $(0.45)45$

Explain your answer.

_____

_____

_____

Unit 7 Guided Practice

Standard 6.3(B) – Supporting

1. Brandy has a new part-time job. On Monday, she worked 8 hours. On Wednesday, she worked three-fourths as long as she did on Monday. Which statement is NOT true about the number of hours Brandy worked on Monday and Wednesday?

   Ⓐ Brandy worked more hours on Monday than she worked on Wednesday.

   Ⓑ Brandy worked fewer than 8 hours on Wednesday.

   Ⓒ Brandy worked 3 more hours on Monday than Wednesday.

   Ⓓ Brandy worked fewer than 16 hours on Monday and Wednesday combined.

2. Brendan set a goal to sell 30 fundraiser cards by Friday. On Monday, he sold $\frac{1}{3}$ of his goal. On Tuesday, he sold $\frac{3}{2}$ of his goal to friends and family. Which of the following most accurately compares the number of cards Brendan sold on Monday and Tuesday to his goal?

   Ⓕ He sold more cards than his goal on Monday, but fewer cards than his goal on Tuesday.

   Ⓖ On both Monday and Tuesday, he sold fewer cards than his goal.

   Ⓗ He sold fewer cards than his goal on Monday, but more cards than his goal on Tuesday.

   Ⓙ On both Monday and Tuesday, he sold more cards than his goal.

3. Stanley drinks $\frac{2}{3}$ cup of milk each day. Which best describes the amount of milk Stanley will drink in 5 days?

   Ⓐ Fewer than $\frac{2}{3}$ cup

   Ⓑ Fewer than 5 cups

   Ⓒ Exactly 5 cups

   Ⓓ More than 5 cups

4. Shanequa began with the number 240. She multiplied the number by $\frac{7}{5}$. Then Shanequa multiplied that product by $\frac{13}{13}$. Which of the following is the most accurate statement?

   Ⓕ Shanequa's number increased with the first multiplication and then remained the same with the second multiplication.

   Ⓖ The second product had a value of exactly 240.

   Ⓗ Shanequa's number decreased with the first multiplication and then increased with the second multiplication.

   Ⓙ Shanequa's number increased with the first multiplication and then increased again with the second multiplication.

5. Jonas' team constructs scale models of robots for a competition. Model A is $\frac{3}{4}$ the size of the original robot, while Model B is $\frac{5}{6}$ the original size. Which of the following most accurately compares the scale models to the original robot?

   Ⓐ Model A and Model B are both larger than the original.

   Ⓑ Model A is larger than the original, while Model B is smaller than the original.

   Ⓒ Model A and Model B are both smaller than the original.

   Ⓓ Model A is smaller than the original, while Model B is larger than the original.

Standard 6.3(B) – Supporting

Unit 7 Independent Practice

1 Ms. Green uses $2\frac{1}{8}$ yards of fabric to make bookmarks for her 24 students. Mr. Ang asks her to make bookmarks for his students. If Mr. Ang has twice as many students as Ms. Green, which best describes the amount of fabric needed to make the bookmarks for his students?

Ⓐ Less than 2 yards

Ⓑ Exactly 2 yards

Ⓒ Greater than 2 yards

Ⓓ Not here

2 Kwan spent 27 minutes talking on the phone on Wednesday. On Thursday, Kwan's phone time was $\frac{9}{10}$ of Wednesday's time. On Friday, Kwan's phone time was $\frac{10}{9}$ of Wednesday's time. Which of the following most accurately compares the times Kwan spent on the phone?

Ⓕ Kwan spent more than 27 minutes talking on the phone both Thursday and Friday.

Ⓖ Kwan spent fewer than 27 minutes talking on the phone both Thursday and Friday.

Ⓗ Kwan spent more than 27 minutes talking on the phone Thursday and less than 27 minutes on Friday.

Ⓙ Kwan spent less than 27 minutes talking on the phone Thursday and more than 27 minutes on Friday.

3 Daniel used $\frac{5}{8}$ of a box of nails to build a birdhouse. A box of nails weighs 16 ounces. Which statement is true about the weight of the nails Daniel used?

Ⓐ Daniel used less than 16 ounces of nails.

Ⓑ Daniel used exactly 1 pound of nails.

Ⓒ Daniel used more than 16 ounces of nails.

Ⓓ Daniel used about 2 ounces of nails.

4 Before working the problem shown below, Jasmine predicted the product would be greater than 8.

$$1\frac{1}{2} \times 8$$

Is Jasmine's prediction correct?

Ⓕ Yes, because she is multiplying 8 by a fraction

Ⓖ No, because she is multiplying 8 by a fraction

Ⓗ Yes, because she is multiplying 8 by a value greater than 1

Ⓙ No, because she is multiplying 8 by a value greater than 1

5 Sam made an initial deposit of $320 to his savings account. In June, he made a deposit that was $\frac{3}{2}$ his initial deposit. He made a third deposit in July that was $\frac{5}{6}$ his initial deposit. Which most accurately compares Sam's deposits in June and July to his initial deposit?

Ⓐ There was an increase in the June deposit and a decrease in the July deposit.

Ⓑ There was a decrease in the June and July deposits.

Ⓒ There was a decrease in the June deposit and an increase in the July deposit.

Ⓓ There was an increase in the June and July deposits.

6 Avery needs to triple her cookie recipe. The original recipe calls for $\frac{3}{4}$ cup of sugar. If Avery only has 3 cups of sugar, will she have enough sugar to triple the cookie recipe?

Ⓕ Yes, because $\frac{3}{4}$ times $\frac{3}{3}$ is equal to $\frac{3}{4}$

Ⓖ No, because multiplying a number by 3 will increase its value

Ⓗ Yes, because $\frac{3}{4}$ times 3 is less than 3

Ⓙ No, because 3 times $\frac{3}{4}$ is greater than 3

Unit 7 Assessment

Name _____

Standard 6.3(B) – Supporting

1 Donut King sells glazed donuts for $6 per dozen. Ellen buys donuts for her office. Which statement is NOT true?

Ⓐ Ellen spends less than $6 for 9 donuts.

Ⓑ Ellen spends more than $6 for $1\frac{1}{4}$ dozen donuts.

Ⓒ Ellen spends exactly $18 for $3\frac{3}{4}$ dozen donuts.

Ⓓ Ellen spends more than $24 for 50 donuts.

2 Arnold's Hot Dog Stand sold $8,014 worth of hot dogs during the month of June. Arnold's July sales were $1\frac{1}{8}$ times as much as his June sales. Arnold's August sales were $\frac{9}{10}$ as much as his June sales. Which of the following is the most accurate statement comparing June sales with the other two months?

Ⓕ Arnold's sales increased in July and increased again in August.

Ⓖ Arnold's sales decreased in July and decreased again in August.

Ⓗ Arnold's sales decreased in July and increased in August.

Ⓙ Arnold's sales increased in July and decreased in August.

3 Jacob's backpack weighs 8 kilograms. Emma's backpack weighs $\frac{2}{3}$ as much as Jacob's backpack. Which is a correct estimate of the weight of Emma's backpack?

Ⓐ Less than $\frac{2}{3}$ kilogram

Ⓑ Greater than 8 kilograms

Ⓒ Exactly 8 kilograms

Ⓓ Less than 8 kilograms

4 Mr. Thom gave his students a riddle. He read the first clue, "When you multiply a whole number by this number, it will decrease the value." Which values could represent Mr. Thom's number?

I. $\frac{3}{8}$

II. $\frac{5}{4}$

III. 0.98

IV. 1.05

Ⓕ II and IV only   Ⓗ I, II, and III only

Ⓖ I and III only   Ⓙ I, II, III, and IV

5 Raj found a coat online for $75, but it was sold out. While shopping at the mall, he found the same coat at two different stores. Fin's was selling the coat for $\frac{4}{5}$ of the online price, while Rainey's price was $\frac{7}{6}$ of the online price. Which store had the better buy on the coat?

Ⓐ Fin's, because the coat costs less than $75

Ⓑ Rainey's, because the coat costs more than $75

Ⓒ Fin's, because the coat costs exactly $45

Ⓓ Rainey's, because the coat costs less than $75

6 Alli uses a copy machine to reproduce a photograph so that it is $\frac{3}{2}$ the size of the original. Which statement is true about the new photograph compared to the original?

Ⓕ The new photograph is smaller because $\frac{3}{2}$ is a fraction.

Ⓖ The new photograph is larger because $\frac{3}{2}$ is greater than one.

Ⓗ The new photograph remains the same size.

Ⓙ You cannot determine how the new photograph compares to the original.

Standard 6.3(B) – Supporting

Unit 7 Critical Thinking

1 Jeanine, Marcus, and Heather practice running for marathons at the local park. Jeanine sets a goal to run 15 miles each week. Marcus sets a weekly goal that is $\frac{5}{3}$ Jeanine's goal. Heather sets a weekly goal that is $\frac{4}{5}$ Marcus' goal. Complete the table, showing the runners in order from greatest to least, according to their weekly goals. Show all work.

	Runner	Weekly Goal (mi)
1		
2		
3		

2 Write and solve a word problem containing the factor 5 and a second factor that is a fraction. The solution must be greater than 5.

Unit 7 Journal/Vocabulary Activity

Standard 6.3(B) – Supporting

Journal

Describe a time when it would be useful to determine if the product of two numbers increases or decreases in value without multiplying the two numbers.

_____
_____
_____

Vocabulary Activity

Circle the word or phase that makes each statement true. Write two equations to verify each statement.

1. Multiplication (always/sometimes/never) increases the value of a quantity.

    Equations: _____

2. When multiplying a quantity by a fraction in which the numerator and denominator are equal, the value of the quantity will (increase/decrease/remain the same).

    Equations: _____

3. When multiplying a quantity by a fraction in which the numerator is larger than the denominator, the value of the quantity will (increase/decrease/remain the same).

    Equations: _____

4. When multiplying a quantity by a fraction in which the numerator is smaller than the denominator, the value of the quantity will (increase/decrease/remain the same).

    Equations: _____

5. The product of two quantities greater than one results in a value (larger than/smaller than/equivalent to) either factor.

    Equations: _____

6. The product of two quantities between zero and one results in a value (larger than/smaller than/equivalent to) either factor.

    Equations: _____

Name _____

Standard 6.3(B) – Supporting

 Unit 7 Motivation Station

On a Roll

Play *On a Roll* with a partner. Each pair of players needs a number cube and a game board. Each player needs a pencil. Player 1 rolls the number cube and solves the corresponding problem in the space provided. If correct, player 1 is awarded the number of points in the banner next to the problem. If incorrect, play passes to player 2. If a player rolls the number of a problem that has already been solved, play passes to the next player. The winner is the player with more points after all problems have been solved.

Is $\frac{2}{3} \times 5$ greater than or less than 5?	Is $\frac{1}{2} \times 2$ greater than or less than $\frac{1}{4} \times 2$?
Kara painted a picture that was 6 inches wide and $\frac{17}{2}$ inches long. Taylor painted a picture the same width but $\frac{11}{2}$ inches long. How do the areas of the pictures compare? 	How does $40 \times \frac{5}{3}$ compare to $\frac{3}{5} \times 40$?
Renee baked a cake in a pan that is $12\frac{3}{4}$ inches long and 6 inches wide. Is the area of the pan larger or smaller than 72 square inches? 	Lauren is carpeting two rooms in her house. She needs 600 square yards of carpet for her living room and $\frac{5}{6}$ of that amount for her bedroom. How does the amount of carpet needed for her bedroom compare to the amount of carpet needed for her living room? 

Unit 7 Homework

Name _____

Standard 6.3(B) – Supporting

1 Alan had two mirrors installed in his house. The first mirror was 2 yards tall and $\frac{2}{3}$ yard wide. The second mirror was 2 yards tall and $\frac{3}{2}$ yards wide.

Is the area of the first mirror greater than or less than 2 square yards? Explain.

_____
_____

Is the area of the second mirror greater than or less than 2 square yards? Explain.

_____
_____

Which mirror has the greater area?

_____

2 Fill in each blank with a different fraction that makes each number sentence true.

6 × _____ > 6

25 < _____ × 25

_____ × 12 < 12

50 > 50 × _____

3 Everett purchases $\frac{7}{8}$ pound of grapes. Grapes are on sale this week for $1.98 per pound. Will the grapes Everett purchases cost more than $1.98 or less than $1.98? Explain.

_____
_____

4 Last week Mrs. Duncan rented a movie that was 99 minutes long. This week she rented a movie that was $\frac{4}{3}$ the length of last week's movie. Did the length of the movie Mrs. Duncan rented this week increase or decrease from last week?

_____

What is the length of the movie, in minutes, that Mrs. Duncan rented this week?

_____

5 Presley ran 5 miles in 38 minutes. Makenna ran the same distance in $\frac{5}{6}$ the time it took Presley. Complete the statement below.

Makenna ran (faster/slower) than Presley.

Explain your answer.

_____
_____

Connections

Remove the face cards from a deck of cards. Turn over 3 of the remaining cards in the deck. The first card represents a whole number, the second card is the numerator of a fraction, and the third card is the denominator of the fraction. Determine whether the product of the whole number and the fraction will be greater than the whole number or less than the whole number. Continue until all the cards have been used.

Standard 6.3(C) – Supporting

Unit 8 Introduction

1 Write an equation that could be represented by the model below.

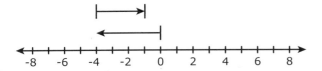

_____

2 Write one division equation and one multiplication equation for the model shown below.

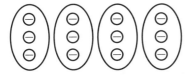

_____
_____

3 At 6:00 A.M., the temperature was -9°F. At noon, the temperature was 10°F. Use the number line to model the change in temperature from 6:00 A.M. to noon.

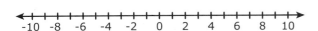

What was the temperature change?

_____

4 Draw two different models that could be used to solve the following equation.

$-3 + (-4) = $ _____

5 Antonio let Octavian borrow $15 at the school book fair to buy 3 books. Octavian paid back $4 the following week. Draw a model and write an expression to represent this situation.

_____

What integer represents the amount Octavian still owes Antonio?

_____

6 Draw a model to represent $-20 \div 4$.

What quotient is shown in your model?

_____

Unit 8 Guided Practice

Standard 6.3(C) – Supporting

1. At the first of the month, Linda had $1,500 in her checking account. She wrote three checks for a total of $750. She made two deposits, one for $375 and one for $1,125. Which model can be used to determine the balance in Linda's checking account at the end of the month?

Ⓐ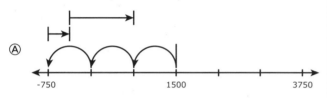

Ⓑ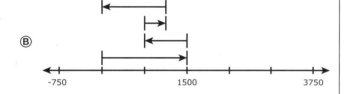

Ⓒ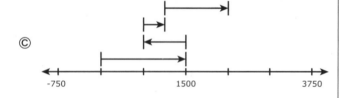

Ⓓ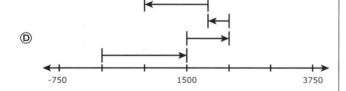

2. Which expression best matches the model shown?

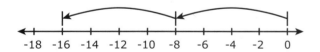

Ⓕ (2)(-8)

Ⓖ 2 + (-8)

Ⓗ (1)(-8)

Ⓙ (1)(-16)

Use the model to answer questions 3 and 4.

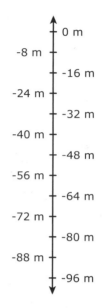

3. A submarine took 16 minutes to reach a depth of 96 meters below the surface of the water. Which integer best represents the rate of the submarine's dive?

Ⓐ 16 meters per minute

Ⓑ -16 meters per minute

Ⓒ 6 meters per minute

Ⓓ -6 meters per minute

4. The submarine took only 12 minutes to come back to the surface again. Which integer best represents the rate of the submarine's ascent?

Ⓕ -8 meters per minute

Ⓖ 8 meters per minute

Ⓗ 12 meters per minute

Ⓙ -12 meters per minute

Standard 6.3(C) – Supporting

Unit 8 Independent Practice

1 Which model best represents 3(-3)?

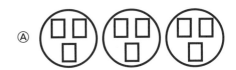
Ⓐ

Ⓑ

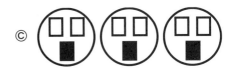
Ⓒ

Ⓓ

2 Which of the following models best represents -24 ÷ 8?

Ⓕ

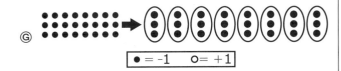
Ⓖ

Ⓗ

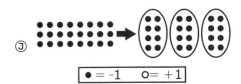
Ⓙ

3 Which of the equations could be used to represent the model shown below?

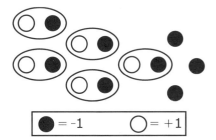

Ⓐ 5 + (-8) = -3

Ⓑ -8 + (-5) = -3

Ⓒ 5 − (-8) = -3

Ⓓ 8 − (-5) = -3

4 Mara opens a checking account. The first four transactions on her account are modeled below.

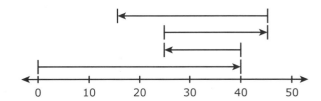

Which best describes the transactions made on Mara's account?

Ⓕ Use debit card for a $40 purchase
   Deposit $15 from babysitting
   Write a check for $20
   Deposit $30 from allowance

Ⓖ Deposit $40 from babysitting
   Use debit card for a $15 purchase
   Deposit $20 from allowance
   Write a check for $30

Ⓗ Deposit $40 from babysitting
   Use debit card for a $25 purchase
   Write a check for $45
   Deposit $15 from allowance

Ⓙ Deposit $40 from babysitting
   Deposit $20 from allowance
   Use debit card for a $15 purchase
   Write a check for $30

Unit 8 Assessment

Standard 6.3(C) – Supporting

1 Which equation best represents the model?

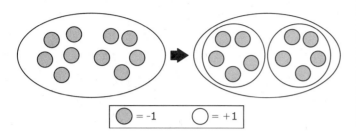

- Ⓐ $-10 \div (-2) = -5$
- Ⓑ $-10 \div 2 = 5$
- Ⓒ $-10 \div 2 = -5$
- Ⓓ $10 \div 2 = 5$

2 Mickey opened a new bank account in January. He kept a record of the activity in his bank account.

Mickey's Bank Account

Date	Deposit	Withdraw
January	$100	$25
February	$50	$40

Which model best represents Mickey's balance at the end of these two months?

Ⓕ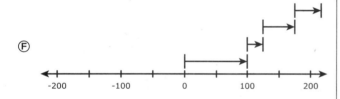

Ⓖ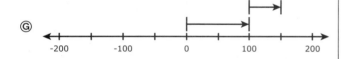

Ⓗ

Ⓙ

3 When Melinda left for school at 8:00 A.M., the temperature was -7°F. When she came home at 4:00 P.M., the temperature was 16°F. Using the number line below as an aid, which equation correctly shows how to find the change in temperature from 8:00 A.M. to 4:00 P.M.?

- Ⓐ $16 - 7 = 9$
- Ⓑ $16 - (-7) = 23$
- Ⓒ $-7 - 16 = -23$
- Ⓓ $16 - (-7) = 9$

4 Jeremy uses tiles to represent integers as shown below.

■ = -1   □ = +1

How would Jeremy represent 2(-4)?

Ⓕ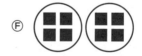

Ⓖ

Ⓗ

Ⓙ

Standard 6.3(C) – Supporting

Unit 8 Critical Thinking

1 Mrs. Vo asked her students to write the following expression using numbers and symbols, and then find the solution.

"Negative five minus negative 2"

Shelly wrote -5 − (-2) = -3. Heraldo wrote -5 + 2 = -3. Which student wrote the equation correctly? _____

Justify your answer using models.

2 Write a word problem that could be solved using the following model.

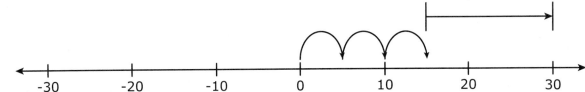

_____
_____
_____
_____
_____

Unit 8 Journal/Vocabulary Activity

Name _____

Standard 6.3(C) – Supporting

Journal

Compare and contrast the process used to add integers with the same sign and the process used to add integers with different signs.

_____

_____

_____

Vocabulary Activity

Use the terms in the box to complete each sentence.

1. Each mark on the number line below indicates an _____.

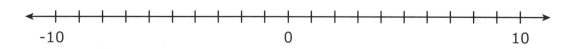

2. The points on the number line below represent _____.

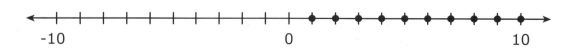

3. The points on the number line below represent _____.

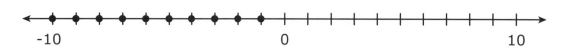

4. The illustration shown on the number line below is an example of a _____.

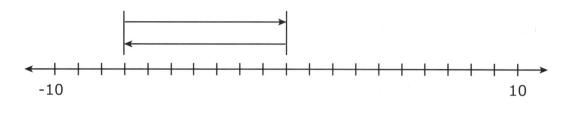

| negative integers | zero pair | integer | positive integers |

Name _____

Standard 6.3(C) – Supporting

Unit 8 Motivation Station

Sketch and Solve

Read each problem on the left page of the notebook. Sketch an integer model to illustrate each problem on the right page of the notebook. Record each solution on the line provided.

Timothy and Anna combine their savings to buy a gift for their grandmother's 90th birthday. The cost of the gift is $41. Timothy has $18, and Anna has $15. How much money do Timothy and Anna still need to buy the gift for their grandmother?

Solution: _____

Julia leaves for school at 7 A.M. and the temperature outside is 38°F. The temperature drops 2° each hour for the next 8 hours until Julia gets out of school. What is the temperature when Julia gets out of school?

Solution: _____

Mark plays his favorite video game. He loses 30 points in 10 minutes. What integer represents Mark's average points per minute?

Solution: _____

Unit 8 Homework

Name _____

Standard 6.3(C) – Supporting

1. Draw a number line model to represent the following problem: -5 + 4 + (-8).

   -5 + 4 + (-8) = _____

2. The model below shows the steps in an integer computation problem.

   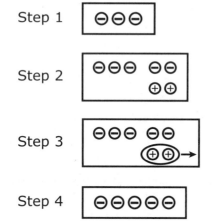

   What number sentence represents the model?

   _____

3. Mr. Garrison purchases a new freezer. Before he can store food in it, the freezer must be cooled to -18°C. The current temperature in the freezer is 21°C. If the temperature in the freezer decreases at a rate of 3° per hour, how long will it take the freezer to reach -18°C? Draw a model to support your solution.

   _____

4. Write an equation to represent the following model.

   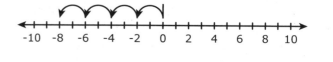

   _____

5. Model and solve the following.

   $$\frac{-15}{3}$$

Connections

Using a standard deck of cards, separate the aces through nines from the tens and face cards. Red number cards (A–9) represent negative numbers while black number cards (A–9) represent positive numbers. The remaining cards represent the following: 10–addition; Jack–subtraction; Queen–multiplication; King–division. Turn over two number cards and an operation card. Draw a model to represent the problem and find the answer. Repeat until all cards have been used.

Standard 6.3(D)

Unit 9 Introduction

1 Simplify the expressions below.

a. -29 − 13

b. $\dfrac{-60}{5}$

c. -4 + 6 × 2

d. -5(-7)

2 Kyle received a notice from his bank that his account was overdrawn by $14. He made a deposit that brought his balance to $50. How much money did Kyle deposit?

3 A submarine travels 330 feet below sea level. If the submarine ascends 125 feet, what integer expresses the submarine's new position?

4 A scuba diver dove 950 feet below the surface in 19 minutes. What integer represents the average rate of change for this dive?

5 The low temperature in Anchorage, Alaska, is -15°F while the low temperature in Lindale, Texas, is 32°F. Write an expression to find the difference between the low temperatures in Lindale and Anchorage.

_____

What is the temperature difference? _____

6 An airplane's position changes -1,500 feet per minute. What integer represents the total change in position of the airplane after 5 minutes?

Unit 9 Guided Practice

Standard 6.3(D)

1. When Bonita started driving at 4:00 P.M., the temperature was 12°F. When she arrived home at 10:00 P.M., the temperature was -5°F. What was the difference in the temperatures at 4:00 P.M. and 10:00 P.M.?

   Ⓐ 7°F

   Ⓑ 17°F

   Ⓒ -5°F

   Ⓓ -7°F

2. An airplane is flying at an altitude of 5,000 feet and starts descending at a rate of 150 feet per minute. What will be the altitude of the plane after 10 minutes?

   Ⓕ 3,500 ft

   Ⓖ 4,850 ft

   Ⓗ 4,840 ft

   Ⓙ 1,500 ft

3. Which of the following expressions has equivalent solutions?

   I. -12(4)

   II. -11 + 59

   III. -24 - 24

   IV. $\frac{-98}{-2}$

   Ⓐ I and IV only

   Ⓑ II and III only

   Ⓒ I and III only

   Ⓓ I, II, and III only

4. Detric's football team lost a total of 56 yards in their first game. Which expression results in an integer that represents the team's average yardage per quarter?

   Ⓕ $\frac{-56}{4}$

   Ⓖ -56 ÷ 25

   Ⓗ -56 ÷ (-4)

   Ⓙ $\frac{56}{-25}$

5. Mr. Champion opened a savings account to begin saving money for Jennie's college fund. He initially deposited $250 in the account. Mr. Champion withdraws $50 each month from his checking account to deposit into Jennie's college fund. How much money will be in Jennie's college fund at the end of two years?

   Ⓐ $600

   Ⓑ $800

   Ⓒ $1,200

   Ⓓ $1,450

6. Which of the following expressions is equivalent to -8?

   I. -32 ÷ (-4)

   II. -3 - (-11)

   III. (2)(-4)

   IV. -14 + 6

   Ⓕ I and II only

   Ⓖ III and IV only

   Ⓗ II, III, and IV only

   Ⓙ I, II, III, and IV

Standard 6.3(D)

Unit 9 Independent Practice

1. A diver at sea level needs to dive to a depth of -2,200 feet in 10 minutes. Which integer best represents the diver's rate?

   Ⓐ -22,000 ft/min

   Ⓑ -220 ft/min

   Ⓒ 22,000 ft/min

   Ⓓ 220 ft/min

2. Which of the following is NOT equivalent to -12?

   I. $\frac{36}{-3}$

   II. -4 − (-16)

   III. (-3)(-4)

   IV. 21 + (-9)

   Ⓕ I only

   Ⓖ II and III only

   Ⓗ II, III, and IV only

   Ⓙ I, II, III, and IV

3. On Wednesday, the temperature dropped an average of 4° per hour. Which integer represents the change in temperature from 10 A.M. to 2 P.M.?

   Ⓐ -20°

   Ⓑ 1°

   Ⓒ -16°

   Ⓓ 12°

4. Choy has $25 in his lunch account at school. He spends $3 each day to eat in the cafeteria. What is Choy's lunch account balance after two weeks of buying his lunch in the cafeteria?

   Ⓕ $19

   Ⓖ $5

   Ⓗ -$19

   Ⓙ -$5

5. Simplify the following expression.

   $$-6 + 5 \cdot 2$$

   Ⓐ 8

   Ⓑ 4

   Ⓒ -2

   Ⓓ -16

6. During the first round of a game, Ravi had a score of -27 points. He earned 35 points in the second round. What was Ravi's score after the second round?

   Record your answer and fill in the bubbles on the grid below. Be sure to use the correct place value.

Unit 9 Assessment

Name _____

Standard 6.3(D)

1. Which of the following expressions simplifies to -1?

   I. -39 - 38

   II. -1 · (-1)

   III. 19 + (-20)

   IV. $\frac{-16}{-16}$

   Ⓐ III only

   Ⓑ II and III only

   Ⓒ I, II, and III only

   Ⓓ I, III, and IV only

2. By noon, the temperature increased 21° from the morning low of -13°F. What was the temperature at noon?

   Ⓕ 21°F

   Ⓖ 8°F

   Ⓗ 34°F

   Ⓙ 12°F

3. Jackson went deep sea diving with some friends. If he descends at a rate of 4 feet per minute, what integer represents Jackson's depth in a quarter of an hour?

   Ⓐ 100

   Ⓑ -60

   Ⓒ -100

   Ⓓ 60

4. Which expression has the least value?

   Ⓕ -7 · (-9)

   Ⓖ 46 - 74

   Ⓗ -91 + 68

   Ⓙ 156 ÷ (-6)

5. A shark swims at a depth of 275 feet below sea level. After sighting a whale, the shark moves 190 feet closer to the surface. What integer represents the shark's new depth?

   Ⓐ 465

   Ⓑ -125

   Ⓒ -85

   Ⓓ 125

6. Marco has a balance of -$19 in his checking account. He makes a trip to the bank to deposit $368. The next day he purchases a new video game for $47 using his debit card. What is Marco's new balance after he purchases the video game?

   Ⓕ $434

   Ⓖ $339

   Ⓗ $302

   Ⓙ $301

Name _____

Standard 6.3(D)

Unit 9 Critical Thinking

1. Mrs. Brown's baby boy weighed 6 pounds 4 ounces at birth. After the first week he lost 11 ounces. In week two he gained $\frac{1}{10}$ of his original birth weight, and in week three his weight decreased by $\frac{1}{4}$ pound. Use integers to write an expression to determine the baby's weight, in ounces, three weeks after he was born.

_____

How much did the baby weigh in pounds and ounces at three weeks old? _____

2. Two integers are added, and the sum is a negative integer. Which of the following could yield this result? Circle your answer(s). Give an example to justify your answer.

   a. positive integer + negative integer _____

   b. positive integer + positive integer _____

   c. negative integer + negative integer _____

3. Create a word problem using the following equation.

$$-2(5) = -10$$

_____
_____
_____
_____

Unit 9 Journal/Vocabulary Activity

Name _____

Standard 6.3(D)

Journal

While working on her daily math warm-up, Megan thought to herself that two of the questions were really the same problem.

Daily Warm-Up
1. 5 − 11
2. -5 + 11
3. 5 + (-11)
4. -5 − 11

Is Megan correct? If yes, which two questions is she thinking about and why are they the same? If no, explain why Megan is not correct.

_____

_____

_____

Vocabulary Activity

Brainstorm words or phases that represent negative numbers, positive numbers, and zero. List the words in the table below.

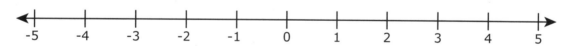

Words that mean NEGATIVE	Words that mean ZERO	Words that mean POSITIVE

Name _____

Standard 6.3(D)

Unit 9 Motivation Station

Walk the Integer Path

Play *Walk the Integer Path* with a partner. Each pair of players needs one game sheet and two paper clips to use with the spinners. Each player needs a small game token or marker and a pencil. Players place their markers at 0. Player 1 spins both spinners to determine the operation and a number. Player 1 then moves his/her marker to reflect the spins. For example, player 1 has a token at 0. If player 1 rolls "+" and "-1," the token is moved to the answer, -1, since 0 + -1 = -1. Play shifts to player 2 who repeats the process. For the next round, perform the correct operation for the number at the new position of the token, and then move the game token to the new answer. The game ends when the teacher calls time. The player with a token at the greater value is the winner.

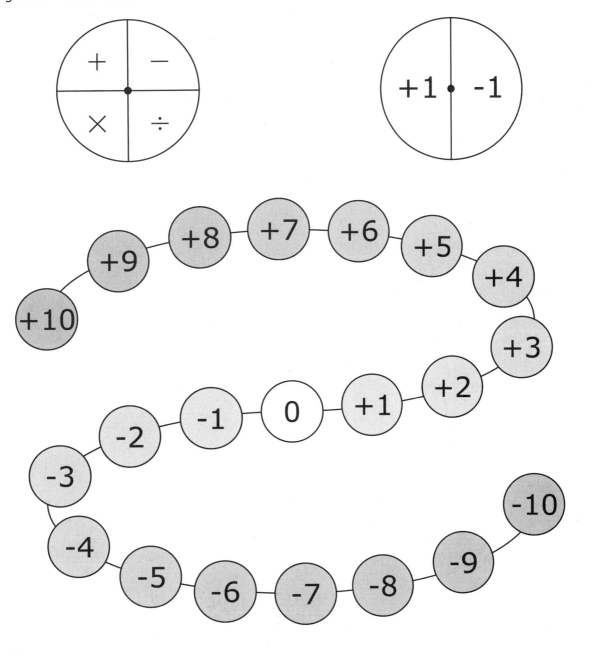

Unit 9 Homework

Standard 6.3(D)

1 Leon scuba dives during his summer vacation. After reaching a depth of 125 feet below sea level, he ascends at a rate of 13 feet per minute. Will Leon reach the surface in 9 minutes? Explain.

_____

_____

_____

2 Michelle's checking account shows the following entries.

Date	Activity	Amount
Aug 10	Balance	$771
Aug 15	Deposit	$250
Aug 17	Withdrawal	$1,028

What is Michelle's account balance after the withdrawal on August 17?

3 Fill in each blank with an integer that will make the equation true.

a. $6 \times$ _____ $= -24$

b. _____ $- 11 = -30$

c. $\dfrac{\text{_____}}{-25} = -4$

d. $9 +$ _____ $= -3$

4 At 5 A.M. the temperature in Bloomington, Minnesota, is $-21°F$. By 11 A.M. the temperature has risen 9°. What is the temperature in Bloomington at 11 A.M.?

5 A submarine travels at a depth of 432 feet below sea level, while an airplane flies at an altitude of 1,388 feet. What is the distance between the submarine and the airplane?

6 In the last three plays of the second quarter, the Belton Bulldogs football team lost 27 yards. Write and solve an equation using integers to determine the team's average yards per play.

Connections

1. Find an example of integers in a newspaper or magazine. Glue the example at the top of a sheet of paper. Write and solve a word problem related to the example.

2. Choose games from the following websites to practice integer computation skills at home:
http://www.mathnook.com/math/skill/integergames.php
http://www.math-play.com/integer-games.html

Name _____

Standard 6.3(E) – Readiness

Unit 10 Introduction

1. Kristin is making spaghetti for a party. Her recipe calls for three-fourths cup of tomato paste. If she needs to make $1\frac{1}{2}$ recipes, how much tomato paste will Kristin use?

2. Eli's paycheck before taxes is $198.65. If he worked 27.4 hours, how much is Eli paid per hour?

3. Hank has a board that is $6\frac{2}{3}$ feet long. He cuts the board into equal pieces that are each $\frac{5}{6}$ foot long. How many $\frac{5}{6}$-foot pieces can Hank cut from the board?

4. The transportation department is resurfacing 12.75 miles of roadways in Navarro County. The crew completes work on a $\frac{1}{8}$-mile section of roadway each hour. How many hours will it take to complete the resurfacing project?

5. Willie found shirts at a back-to-school sale for $15.25 each. He also found jeans on sale for $32.95 per pair. He purchased 3 shirts and 2 pairs of jeans. How much money did Willie spend on the shirts and jeans before tax?

6. Salty Pete's Saltwater Taffy Shop sold 86.4 kilograms of saltwater taffy in the first 6 days of July. What was the average amount of taffy sold each day?

    If each kilogram of taffy sold for $8.15, how much money did Salty Pete's make each day?

©2014 mentoringminds.com   motivationmath™ LEVEL 6   ILLEGAL TO COPY   79

Unit 10 Guided Practice

Name _____

Standard 6.3(E) – Readiness

1. Caroline opens a savings account at her local bank. She plans to put $\frac{1}{10}$ of her net paycheck in her savings account every week. In June, her paychecks are $164.69, $177.86, $187.74, and $163.04. How much money did Caroline put in her savings account in June?

   Ⓐ $6,933.30
   Ⓑ $6.94
   Ⓒ $69.33
   Ⓓ $693.34

2. A cookie cake is cut into 8 equal slices. Damian eats $\frac{1}{8}$ of the cake, Shontavia eats $\frac{1}{2}$ of the cake, and Michael eats 2 slices. How many slices of cake remain?

   Ⓕ 1 slice
   Ⓖ 3 slices
   Ⓗ 5 slices
   Ⓙ 7 slices

3. Davis wants to run across the country. The distance from San Francisco to Albany is 4,108.72 kilometers. Davis runs 5 hours each day at an average speed of 8 kilometers per hour. Which is closest to the number of days it will take Davis to run from San Francisco to Albany?

   Ⓐ 514 days
   Ⓑ 122 days
   Ⓒ 225 days
   Ⓓ 103 days

4. A Mighty Kids frozen meal includes 3.2 ounces of applesauce. If a child eats two frozen meals each week, how many ounces of applesauce will they consume in one year?

   Ⓕ 6.4 oz
   Ⓖ 25.6 oz
   Ⓗ 332.8 oz
   Ⓙ 3,328 oz

5. Mandissa and Katelynn each make a pitcher of strawberry lemonade. Mandissa wants her drink to be more strawberry flavored with less lemon. She uses $2\frac{1}{4}$ times as many strawberries as Katelynn uses, and $\frac{2}{3}$ the amount of lemons. If Katelynn uses $\frac{3}{4}$ cup of strawberries and 6 lemons to make her pitcher of lemonade, how many strawberries and lemons will Mandissa use?

   Ⓐ 3 cups of strawberries and $6\frac{2}{3}$ lemons
   Ⓑ $1\frac{11}{16}$ cups of strawberries and 4 lemons
   Ⓒ 3 cups of strawberries and 4 lemons
   Ⓓ $1\frac{11}{16}$ cups of strawberries and $6\frac{2}{3}$ lemons

6. Carmey and Betty are wrapping gifts for their mother. Carmey is making the bows. She has $37\frac{1}{3}$ yards of ribbon. If each bow uses $4\frac{2}{3}$ yards of ribbon, how many bows can Carmey make?

   Ⓕ 7 bows
   Ⓖ 8 bows
   Ⓗ 9 bows
   Ⓙ Not here

Standard 6.3(E) – Readiness

Unit 10 Independent Practice

1 On Monday, 3.6 centimeters of rain fell in 4.5 hours. What was the average amount of rainfall per hour?

Ⓐ 0.71 cm

Ⓑ 0.8 cm

Ⓒ 1.25 cm

Ⓓ Not here

2 It took Mort one hour to ride $5\frac{3}{4}$ miles on his bicycle. How far will Mort be able to ride in $3\frac{1}{2}$ hours?

Ⓕ $20\frac{1}{8}$ mi

Ⓖ $9\frac{1}{4}$ mi

Ⓗ $1\frac{1}{4}$ mi

Ⓙ $15\frac{3}{8}$ mi

3 Angela uses $7\frac{1}{2}$ inches each of 6 different colors of embroidery floss to make one friendship bracelet. Each package of floss contains 26 feet. How many friendship bracelets can Angela make if she has one package of each color?

Ⓐ 249 bracelets

Ⓑ 41 bracelets

Ⓒ 39 bracelets

Ⓓ 3 bracelets

4 Jacob and his three friends went out to eat. They each ordered an individual pizza and a drink for $6.19. They also ordered breadsticks to share for $4.25. The tax on their bill was $2.39, and they each put in $1.50 for a tip. How much did each person spend on the meal?

Ⓕ $12.47

Ⓖ $8.23

Ⓗ $9.35

Ⓙ $10.97

5 Raul's car averages 17.3 miles per gallon of gasoline. How many miles can Raul drive if he fills his tank with 10.5 gallons of gasoline?

Record your answer and fill in the bubbles on the grid below. Be sure to use the correct place value.

6 The Goode family built a rectangular swimming pool in their backyard. The floor of the pool has an area of $485\frac{5}{8}$ square feet. If the width of the pool is $18\frac{1}{2}$ feet, what is the length of the pool?

Ⓕ $13\frac{1}{8}$ ft

Ⓖ $224\frac{5}{16}$ ft

Ⓗ $8,984\frac{1}{16}$ ft

Ⓙ $26\frac{1}{4}$ ft

Unit 10 Assessment

Standard 6.3(E) – Readiness

1. Davis has 141 figures in his action figure collection. One-third of the figures are sports related. How many of the figures in Davis' collection are NOT sports related?

    Ⓐ 47
    Ⓑ 94
    Ⓒ 423
    Ⓓ 59

2. The diagram below shows a wall with a window in Alejandra's room.

    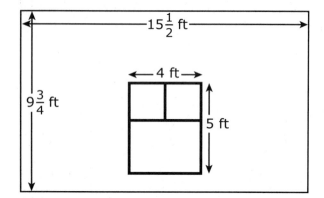

    Alejandra plans to cover the wall with wallpaper. How many square feet of wall paper will she need?

    Ⓕ 151.125 ft²
    Ⓖ 147 ft²
    Ⓗ 131.125 ft²
    Ⓙ 135 ft²

3. Rebecca bought 1.5 pounds of grapes and paid $3.72. What was the cost per pound of the grapes?

    Ⓐ $5.58
    Ⓑ $2.48
    Ⓒ $2.22
    Ⓓ Not here

4. In $3\frac{3}{4}$ hours one day in January, it snowed 3 feet in Amarillo. How much snow fell per hour?

    Ⓕ $\frac{4}{5}$ ft
    Ⓖ $1\frac{1}{4}$ ft
    Ⓗ $11\frac{1}{4}$ ft
    Ⓙ $1\frac{1}{2}$ ft

5. Laurie uses 0.25 pounds of butter to make a batch of her grandmother's cookies. She needs to make 4.5 times the recipe for the school bake sale, and an additional two batches for her family. How much butter does Laurie use to make all the cookies?

    Ⓐ 1.125 lb
    Ⓑ 3.125 lb
    Ⓒ 2.25 lb
    Ⓓ 1.625 lb

Name _____

Standard 6.3(E) – Readiness

Unit 10 Critical Thinking

1 Jayme bought two 20-oz bottles of soda for $2.22. Sam bought a six-pack of 16.9-oz bottles for $2.50. Who got the better deal on the sodas? Use words and numbers to justify your answer.

2 Complete the following computations.

$$16\frac{7}{8} \div 1\frac{1}{4} \qquad\qquad 1.25\overline{)16.875}$$

Do you think it is easier to divide with fractions or decimal numbers? Explain your reasoning.

_____

_____

_____

©2014 mentoringminds.com   motivationmath™ LEVEL 6   ILLEGAL TO COPY   83

Unit 10 Journal/Vocabulary Activity

Name _____

Standard 6.3(E) – Readiness

Journal

Explain how a remainder in a division problem is related to a fraction.

_____
_____
_____

Vocabulary Activity

Sort the following terms or representations into the correct spaces on the chart.

$b\overline{)a}$	$\frac{a}{b}$, where $b \neq 0$	the number of b groups in a	factor
a groups of b	product	divisor	$(a)(b)$	ab
remainder	a times as much as b	$a \cdot b$	$a \div b$
a grouped into b number of groups		dividend	quotient

Multiplication	Division
Symbolic Representations	Symbolic Representations
Verbal Representations	Verbal Representations
Vocabulary Terms	Vocabulary Terms

Standard 6.3(E) – Readiness

Unit 10 Motivation Station

What's Your Story?

Create story contexts for the expressions below. Solve the story contexts in the work space provided. Then, write the solution in a complete sentence in the last column. Share your problems and solutions with the class.

Story	Work Space	Solution
$2\frac{1}{2} \div 4$		
$1.8 \cdot 0.5$		
$10.75 \div \frac{1}{4}$		
$(1\frac{2}{3} \times 3) \div \frac{1}{2}$		

Unit 10 Homework

Standard 6.3(E) – Readiness

1. Humberto earns $7.25 per hour. If he works 22.4 hours each week, how much money will Humberto earn in one year?

2. Stephen makes wooden shelves for his room. He starts with a board that is 20 feet long. Stephen wants to make the shelves $\frac{3}{4}$ foot long. How many shelves can Stephen make?

3. Antoine has $65.72 in his savings account. After 3 months, his bank adds $\frac{1}{100}$ of his current balance to his account in interest. What is Antoine's account balance after 3 months?

4. Leah's mom bought a bag of pistachios. According to the bag, one serving of pistachios is $\frac{2}{3}$ cup. If the bag contains $11\frac{5}{8}$ servings, how many cups of pistachios are in the bag?

5. Julio has $3.30 left from his allowance. He buys pencils that cost $0.55 each. How many pencils can Julio buy?

Connections

1. Using a favorite recipe, determine how the measurements will change if the recipe is cut in half, thirds, or doubled. As a bonus, calculate the amount of each ingredient needed if the recipe is adjusted to make 100 servings.

2. When shopping at the grocery store, use a calculator to determine the price per ounce for different items. Use this information to determine which brand or size is the better buy. Find the total price for buying multiple items or multiple pounds of an item.

Name _____

Standard 6.4(A) – Supporting

Unit 11 Introduction

1 The table below is labeled with the rule, $y = ax$. Complete the table for $a = 3$.

$y = ax$

x	y
-2	
-1	
0	
1	
2	

Graph the ordered pairs to form a line.

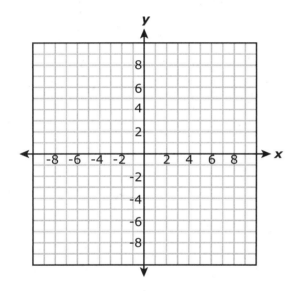

Write an equation to represent the rule shown in the table.

_____

Circle the word that best describes the relationship shown in the table and the graph.

      Additive   Multiplicative

Does the graph intersect the origin? _____

If the value of a changes to -3, will the new graph intersect the origin? _____

Explain. _____

_____

2 The table below is labeled with the rule, $y = x + a$. Complete the table for $a = 3$.

$y = x + a$

x	y
-2	
-1	
0	
1	
2	

Graph the ordered pairs to form a line.

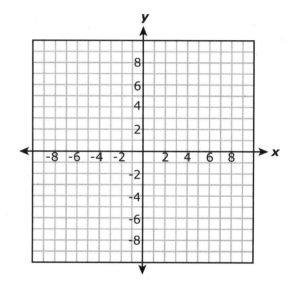

Write an equation to represent the rule shown in the table.

_____

Circle the word that best describes the relationship shown in the table and the graph.

      Additive   Multiplicative

Does the graph intersect the origin? _____

If the value of a changes to -3, will the new graph intersect the origin? _____

Explain. _____

_____

Unit 11 Guided Practice

Standard 6.4(A) – Supporting

1 Which of the following represents a multiplicative relationship?

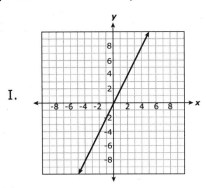
I.

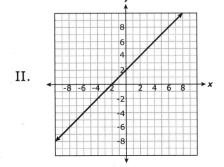
II.

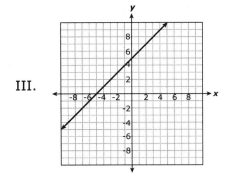
III.

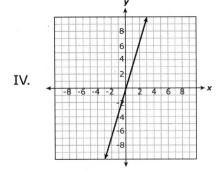
IV.

Ⓐ II and III only

Ⓑ I and IV only

Ⓒ I, II, and IV only

Ⓓ I, II, III, and IV

2 Which statement is NOT true?

Ⓕ Additive relationships are represented using the equation $y = x + a$.

Ⓖ The graph of $y = ax$ will always intersect the origin.

Ⓗ Multiplicative relationships are represented using the equation $y = ax$.

Ⓙ The graph of $y = x + a$ will always intersect the origin.

3 Which statement is true about the tables?

Table 1

x	y
0	0
2	6
4	12
6	18

Table 2

x	y
0	4
2	6
4	8
6	10

Ⓐ Both show an additive relationship.

Ⓑ Table 1 shows an additive relationship; Table 2 shows a multiplicative relationship.

Ⓒ Table 1 shows a multiplicative relationship; Table 2 shows an additive relationship.

Ⓓ Both show a multiplicative relationship.

4 Study the situations shown below.

  I. Mya is 8 years older than Simon.

  II. Mya receives $8 each hour she works.

Which statement is true?

Ⓕ Situation I shows a multiplicative relationship, and situation II shows an additive relationship.

Ⓖ Situation I shows an additive relationship, and situation II shows a multiplicative relationship.

Ⓗ Both show multiplicative relationships.

Ⓙ Both show additive relationships.

Standard 6.4(A) — Supporting

Unit 11 Independent Practice

1 Which of the following does NOT show an additive relationship?

Ⓐ
x	y
-5	-3
-3	-1
2	4
4	6

Ⓑ $y = x + 8$

Ⓒ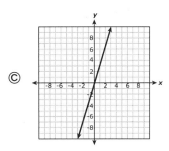

Ⓓ Amy is three years older than Sam.

2 Mrs. McMichael gave all her students 5 extra points on the math test. Instead of 2 points each, Mr. Pringle decided that each correct answer would have a value of 2.5 points. Which statement is true about the equations used to determine the students' test grades?

Ⓕ Mrs. McMichael determined test grades using an additive relationship, $y = x + 5$, where x is the original test score.

Ⓖ Mr. Pringle determined test grades using a multiplicative relationship, $y = 2x$, where x is the number of correct answers.

Ⓗ Mrs. McMichael determined test grades using a multiplicative relationship, $y = 5x$ where x is the number of correct answers.

Ⓙ Mr. Pringle determined test grades using an additive relationship, $y = x + 2.5$, where x is the original test score.

3 Irving and Alex each drew a line on a coordinate plane and then chose two points from their line. Irving's line can be represented by the points (-2, 12) and (7, 21). Alex chose (0, 0) and (3, 15) to represent his linear relationship. Which statement is true about the lines Irving and Alex graphed?

Ⓐ Both Irving's graph and Alex's graph represent a multiplicative relationship.

Ⓑ Irving's graph represents a multiplicative relationship, and Alex's graph represents an additive relationship.

Ⓒ Both Irving's graph and Alex's graph represent an additive relationship.

Ⓓ Irving's graph represents an additive relationship, and Alex's graph represents a multiplicative relationship.

4 Which coordinate plane shows two lines that both represent a multiplicative relationship?

Ⓕ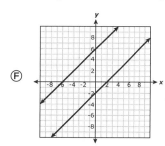

Ⓖ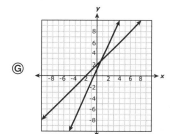

Ⓗ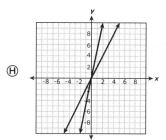

Ⓙ None of the above

Unit 11 Assessment

Standard 6.4(A) – Supporting

1. Israel and Myla create rules using a function machine. Israel uses the equation $y = \frac{5}{2}x$ to describe his rule. Myla uses the equation $y = \frac{2}{5}x$ to describe her rule. Which statement does NOT correctly describe the rules created by Israel and Myla?

   Ⓐ Israel's rule shows a multiplicative relationship.

   Ⓑ The graph of Myla's rule intersects the origin.

   Ⓒ Myla's rule is described by the equation $y = ax$ where a is 0.4.

   Ⓓ The graph of Israel's rule does not intersect the origin.

2. Dawson created four data tables for different numbers of apples and oranges. Which tables show an additive relationship?

   I.
Apples	1	2	3	4
Oranges	5	6	7	8

   II.
Apples	5	10	15	30
Oranges	10	20	30	60

   III.
Apples	7	10	13	16
Oranges	10	13	16	19

   IV.
Apples	1	2	3	4
Oranges	3	6	9	12

   Ⓕ I and II only

   Ⓖ I and III only

   Ⓗ I, II, and III only

   Ⓙ I, III, and IV only

3. Jazmine and Jimmy each created a graph in math class as shown below.

   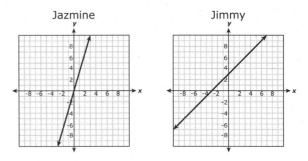

   Which statement is NOT true?

   Ⓐ Jazmine's graph intersects the origin and represents a multiplicative relationship.

   Ⓑ The equation for Jimmy's graph is in the form $y = x + a$.

   Ⓒ The equation for Jazmine's graph is in the form $y = x + a$.

   Ⓓ Jimmy's graph does not intersect the origin and represents an additive relationship.

4. Hannah and Gabby are training for a marathon. They begin the first day running 2 miles. After the first day, Hannah increases her running distance by 1 mile each day. Gabby increases her running distance by $1\frac{1}{2}$ times the previous day. Which statement is true?

   Ⓕ Hannah's training plan shows an additive relationship, while Gabby's training plan shows a multiplicative relationship.

   Ⓖ Both Hannah's and Gabby's training plans show an additive relationship.

   Ⓗ Hannah's training plan shows a multiplicative relationship while Gabby's training plan shows an additive relationship.

   Ⓙ Both Hannah's and Gabby's training plans show a multiplicative relationship.

Name _____

Standard 6.4(A) – Supporting

Unit 11 Critical Thinking

Choose a value for *a* that is between –5 and 5. Use the value selected to write two equations, one multiplicative and one additive. Then, complete the table and plot the two sets of coordinate pairs on the graph.

Equation ($y = ax$): _____

Equation ($y = x + a$): _____

x	Line 1 y = ax	Line 2 y = x + a
-2		
-1		
0		
1		
2		

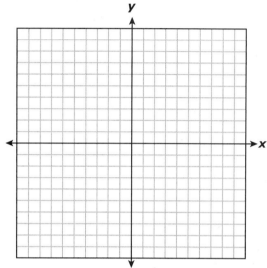

Use the Venn diagram to compare and contrast the two equations and graphs shown above. Be sure to include direction, steepness of the line, whether the line passes through the origin, patterns, etc.

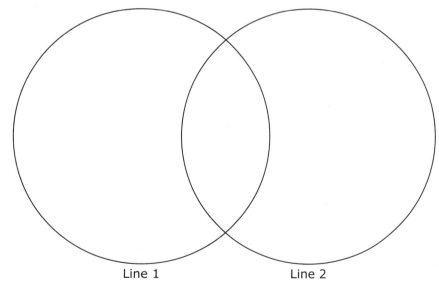

Line 1   Line 2

Partner with another student who selected a different value for *a* and compare your results. How are your graphs the same? How are they different? Explain why you think there are differences in the graphs.

_____
_____
_____

Unit 11 Journal/Vocabulary Activity

Standard 6.4(A) – Supporting

Journal

Give a real-world example of a multiplicative relationship. Explain how you know the relationship is multiplicative.

_____

_____

_____

Vocabulary Activity

For each statement, circle the correct response.

1. The *coefficient* of a term in an expression is added/multiplied with one or more variables.

2. The *constant* term in an expression or equation is a fixed/changing value and does/does not contain variables.

3. Which form of a linear equation is *multiplicative*?

    $y = ax$ $y = x + a$

4. Which form of a linear equation is *additive*?

    $y = ax$ $y = x + a$

5. The *equation* of which graph will always pass through the *origin*?

    $y = ax$ $y = x + a$

6. Put an X on the table and/or graph that shows a *multiplicative* relationship. Circle the table and/or graph that shows an *additive* relationship.

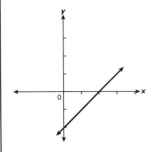

x	y
3	$\frac{9}{4}$
5	$\frac{15}{4}$
7	$\frac{21}{4}$

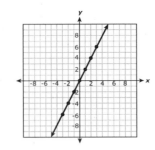

x	y
2	4
4	6
6	8

Name _____

Standard 6.4(A) – Supporting

Unit 11 Motivation Station

So What's Next?

Each set of values for *x* and *y* below is determined by a rule. Three ordered pairs are given. From the answer bank at the bottom of the page, select a fourth set of ordered pairs that follows the same rule. Then, write the rule and state whether it is multiplicative or additive. The first one is done for you.

	Set of Values	Fourth Value	Rule	Multiplicative/ Additive
1.	(2, 3) (4, 6) (6, 9)	(8, 12)	$y = \frac{3}{2}x$	Multiplicative
2.	(1, 4) (2, 8) (3, 12)			
3.	$(3, \frac{9}{2})$ $(5, \frac{13}{2})$ $(7, \frac{17}{2})$			
4.	(2, 8) (3, 9) (4, 10)			
5.	$(5, \frac{5}{2})$ $(7, \frac{7}{2})$ $(9, \frac{9}{2})$			
6.	$(1, \frac{7}{4})$ $(4, \frac{19}{4})$ $(7, \frac{31}{4})$			
7.	(3, 1) (6, 2) (9, 3)			
8.	(4, 6.25) (6, 8.25) (8, 10.25)			
9.	(5, 6) (8, 9.6) (11, 13.2)			

$(11, \frac{11}{2})$ (4, 16)   (14, 16.8) $(9, \frac{21}{2})$ (12, 4) $(10, \frac{43}{4})$ (10, 12.25)   (5, 11)

Explain how you determined the rule for each set of values. _____

_____

Explain how you know if a rule is multiplicative or additive. _____

_____

Unit 11 Homework

Standard 6.4(A) — Supporting

Read the following scenarios and answer the questions that follow.

1 John is renting a kayak for his weekend trip to the lake. The cost to rent the kayak is $15 per day.

Create a table of values for scenario 1. Start with $x = 0$.

Number of Days (x)	Total Cost (y)
0	

What is the rule for scenario 1?

_____

At what point would the graph for this data intersect the y-axis?

_____

What information does that provide about the relationship in scenario 1?

_____

2 Samantha received $3 from her grandmother for her birthday. She put the money in a piggy bank. Every month, she adds $1 to the bank from her allowance.

Create a table of values for scenario 2. Start with $x = 0$.

Number of Months (x)	Total Saved (y)
0	

What is the rule for scenario 2?

_____

At what point would the graph for this data intersect the y-axis?

_____

What information does that provide about the relationship in scenario 2?

_____

Connections

1. Look in a science textbook to find examples of multiplicative and additive relationships. What kind of science topics are involved in each of these relationships? Share your findings with classmates.

2. Write a rule that is both multiplicative and additive. How does your rule differ from the rules you have been studying? How is it the same? Bring your rule to class and share. Can you think of a real-world scenario that would match your rule? Compare with other students in your class.

Name _____

Standard 6.4(B) – Readiness

Unit 12 Introduction

1 William purchases $\frac{1}{2}$ dozen cookies for $5.88. Explain how you can determine the unit cost of the cookies.

_____
_____

What is the unit cost? _____

Write an expression using the unit cost to find the cost of three dozen cookies.

_____
_____

What is the cost of three dozen cookies?

_____
_____

2 Maleeka needs to purchase barbeque sauce for her backyard barbeque. Using the information in the table, explain how Maleeka can determine which sauce is the best buy.

Barbeque Sauce

Sauce Brand	Cost
A	28 oz for $5.04
B	18 oz for $2.88
C	24 oz for $3.60
D	12 oz for $2.04

_____
_____

3 The table shows the results of a survey of 100 people who purchased chips on Saturday.

Chip Survey

Type of Chip	Number of People
Cheese puffs	17
Tortilla chips	26
Potato chips	14
Hot fries	43

If 250 people purchase chips, how many would be expected to buy potato chips?

_____
_____

4 Mrs. Espinola searched the Internet for a punch recipe. She found two similar recipes as shown in the table below.

Recipe	Raspberry Sherbet	Lemon-Lime Soda
Pink Party Punch	3 quarts	6 liters
Raspberry Punch	2 quarts	5 liters

Use the information in the table to complete each statement about the two recipes.

The _____ recipe has a stronger lemon-lime flavor because

_____
_____
_____ .

The _____ recipe has a stronger raspberry flavor because

_____
_____
_____ .

Mrs. Espinola decides to use the Raspberry Punch recipe, but adds an extra quart of raspberry sherbet. Do the Raspberry Punch and the Pink Party Punch now have the same level of raspberry flavor? Explain your answer.

_____
_____
_____

What one change could be made to the original Raspberry Punch recipe so that both punch recipes have an equal ratio of raspberry and lemon-lime flavors?

_____
_____

Unit 12 Guided Practice

Standard 6.4(B) – Readiness

1. Yelena writes a 1,500 word essay about the American Revolutionary War. She writes an average of 350 words each day she works on the essay. What is a reasonable conclusion based on the given information?

   Ⓐ It takes Yelena over 5 days to write the essay.

   Ⓑ Yelena completes almost half the essay by day two.

   Ⓒ Yelena completes exactly one-fourth of the essay on her first day of writing.

   Ⓓ It will only take Yelena seven hours to complete writing the essay.

2. The table below shows the numbers of pencils and pens three friends have in their backpacks.

Name	Number of Pencils	Number of Pens
Lydia	10	7
Angel	9	6
Elizabeth	12	9

   Who has a greater ratio of pencils to pens?

   Ⓕ Lydia

   Ⓖ Angel

   Ⓗ Elizabeth

   Ⓙ The ratio for all three friends is the same.

3. After restocking the vending machine, Efrain determined that 2 out of every 5 packages of candy sold were M&M'S®. The vending machine sold 160 packages of candy last week. How many packages of M&M'S® were sold?

   Ⓐ 400, because $\frac{2}{5} = \frac{160}{400}$

   Ⓑ 16, because $160 \div (5 \times 2) = 16$

   Ⓒ 64, because $\frac{2}{5} = \frac{64}{160}$

   Ⓓ 34, because $\frac{160}{5} + 2 = 34$

4. Mrs. Cavanaugh's chocolate chip peanut butter cookie recipe calls for 2 cups of peanut butter for every $\frac{1}{2}$ cup of chocolate chips. Mrs. Fox's recipe calls for $1\frac{1}{2}$ cups of peanut butter for every $\frac{1}{2}$ cup of chocolate chips. Which statement is true about Mrs. Cavanaugh's and Mrs. Fox's recipes?

   Ⓕ Mrs. Fox's cookies are more "chocolatey" than Mrs. Cavanaugh's cookies.

   Ⓖ Mrs. Cavanaugh's cookies are more "chocolatey" than Mrs. Fox's cookies.

   Ⓗ Both Mrs. Fox's and Mrs. Cavanaugh's cookies contain an equivalent ratio of peanut butter to chocolate chips.

   Ⓙ There is not enough information to compare Mrs. Fox's and Mrs. Cavanaugh's cookies.

5. Mrs. Gregory needs to buy backpacks for the 6 children in her family. She calls 4 stores and records their prices in the table below.

   Backpack Prices

Store	Price
A	6 for $119.70
B	3 for $60.75
C	2 for $39.50
D	$21.99 each

   Mrs. Gregory goes to the store where she will save the most money. From which store does Mrs. Gregory purchase the backpacks?

   Ⓐ Store A, because she needs 6 backpacks

   Ⓑ Store B, because each backpack costs $20.25

   Ⓒ Store C, because the unit cost for each backpack is the least

   Ⓓ Store D, because one backpack costs $21.99

Standard 6.4(B) – Readiness

Unit 12 Independent Practice

1. Misty wants to save money to purchase a new bicycle that costs $279. She opens a savings account and deposits $7 per week. Based on this information, which of the following is a reasonable conclusion?

   Ⓐ Misty will save more than $\frac{1}{2}$ the cost of the bicycle in 20 weeks.

   Ⓑ Misty will save less than $100 after 15 weeks.

   Ⓒ Misty will save less than $\frac{1}{3}$ the cost of the bicycle in 14 weeks.

   Ⓓ Misty will save more than $150 in 21 weeks.

2. Georgia mixes 2 tablespoons of yellow paint and 3 tablespoons of blue paint to make green paint. Braylee makes green paint by mixing 3 tablespoons of yellow paint and $4\frac{1}{2}$ tablespoons of blue paint. Which statement is true about the green paints mixed by Georgia and Braylee?

   Ⓕ Braylee's green paint is more yellow than Georgia's green paint.

   Ⓖ Georgia's green paint is more blue than Braylee's green paint.

   Ⓗ Both Braylee's and Georgia's green paint mixtures contain an equivalent ratio of yellow and blue paint.

   Ⓙ It is not possible to compare the green paints mixed by Braylee and Georgia.

3. Samuel purchases a half-dozen cupcakes for $4.25. Which expression could be used to determine the price for 34 cupcakes?

   Ⓐ 4.25 ÷ 0.5 × 34

   Ⓑ 4.25 ÷ 6 × 34

   Ⓒ 6 ÷ 4.25 × 34

   Ⓓ 0.5 ÷ 4.25 × 34

4. A local grocery store is promoting a sale on four brands of cereal. Which of the following is the best buy?

   BRAND W – $3.98    BRAND X – $2.52
   20 oz   14 oz

   BRAND Y – $4.80    BRAND Z – $3.92
   32 oz   28 oz

   Ⓕ Brand W, because the cost per ounce is $0.199

   Ⓖ Brand X, because it is the least expensive

   Ⓗ Brand Y, because the largest box is always the best buy

   Ⓙ Brand Z, because the unit cost is the least

5. Three neighbors each decide to plant a garden of petunias and daffodils. The table below shows the quantity of flowers in each garden.

Neighbor	Number of Petunias	Number of Daffodils
Mrs. Haskell	48	16
Mr. Myer	84	24
Mrs. Peacock	30	8

   Who has a greater ratio of petunias to daffodils in their garden?

   Ⓐ Mrs. Haskell   Ⓒ Mrs. Peacock

   Ⓑ Mr. Myer   Ⓓ The ratios for all three neighbors are the same.

Unit 12 Assessment

Standard 6.4(B) – Readiness

1. Mosab plans a trip to visit his grandparents. The online directions show four possible routes. He wants to get there as quickly as possible. Which route should Mosab take?

Route	Miles	Average Speed
1	143	67 mph
2	130	65 mph
3	126	56 mph
4	125	50 mph

Ⓐ Route 1, because the average speed is the fastest

Ⓑ Route 2, because it takes 2 hours

Ⓒ Route 3, because it takes the least amount of time

Ⓓ Route 4, because the distance is the shortest

2. The table shows the time spent reading a novel and the number of pages read by each of four friends.

Reading Results

Name	Number of Pages	Amount of Time
Trey	44	110 minutes
Luis	63	1.5 hours
Ethan	31	62 minutes
Roderick	78	2 hours 10 minutes

Which friend reads at the slowest rate?

Ⓕ Trey

Ⓖ Luis

Ⓗ Ethan

Ⓙ Roderick

3. Betty has an order for 160 hair bows. She can make 16 hair bows per day. Which is a reasonable conclusion?

Ⓐ Betty completes $\frac{3}{4}$ of the order in 1 week.

Ⓑ After 3 days of work, Betty completes $\frac{1}{4}$ of the order.

Ⓒ Exactly half of the hair bows will be completed in 6 days.

Ⓓ Betty makes $\frac{1}{4}$ of the bows for her order in $2\frac{1}{2}$ days.

4. Victor purchases 3 movie tickets for $18.75 at the Movie Hut. Which theater sells tickets for the same price per ticket?

Movie Tickets

Theater	Number of Tickets	Total Cost
Star Cinema	5	$31.75
Mega Movie Plex	6	$36.90
Monroe Theater	7	$43.75
Dave's Drive-In	8	$49.60

Ⓕ Star Cinema

Ⓗ Monroe Theater

Ⓖ Mega Movie Plex

Ⓙ Dave's Drive-In

5. Mrs. Yearty makes a saltwater solution using a 2 to 1 ratio of water to salt. Mr. Yearty suggested a water to salt ratio of 3 to 2. Which statement is true?

Ⓐ Mrs. Yearty's saltwater is more salty than Mr. Yearty's saltwater.

Ⓑ Mr. Yearty's saltwater is more salty than Mrs. Yearty's saltwater.

Ⓒ Both Mr. and Mrs. Yearty's saltwater solutions contain an equal water to salt ratio.

Ⓓ The two saltwater solutions cannot be compared.

Name _____

Standard 6.4(B) – Readiness

Unit 12 Critical Thinking

Below is a partial sample of a student group's work with ratios, making an orange drink with different amounts of orange concentrate and water.

Team 3

Mix A

$\frac{2\text{ c concentrate}}{3\text{ c water}} = \boxed{\frac{1\text{ c concentrate}}{1.5\text{ c water}}}$

Mix B

$\frac{5\text{ c concentrate}}{9\text{ c water}} = \boxed{\frac{1\text{ c concentrate}}{1.8\text{ c water}}}$

Mix C

$\frac{1\text{ c concentrate}}{2\text{ c water}} = \boxed{\frac{1\text{ c concentrate}}{2\text{ c water}}}$

Mix D

$\frac{3\text{ c concentrate}}{5\text{ c water}} = \boxed{\frac{1\text{ c concentrate}}{1\frac{2}{3}\text{ c water}}}$

A.

B.

Is the mix with the most water the least orange in flavor? Explain. _____
_____

What makes a mix taste more or less orange? _____
_____

The missing information on part A is, "Which mix has the most orange flavor and why?" What should the students write for part A? _____
_____

The missing information on part B is, "Which mix has the least orange flavor and why?" What should the students write for part B? _____
_____

The team decides to make a batch of their orange drink using Mix D, because they think it tastes best. There are 24 students in the class, and the team wants to give each student $\frac{1}{2}$ cup of drink. How much orange concentrate and water, in cups, should the team use to make their drink? Show all work.

Unit 12 Journal/Vocabulary Activity

Name _____

Standard 6.4(B) – Readiness

Journal

Explain how understanding unit rates can make you a better shopper. Give an example.

_____
_____
_____

Vocabulary Activity

Find the vocabulary words defined below.

Definitions

- A ratio that compares quantities measured in different units
  _____
- A comparison of two or more quantities
  _____
- A comparison of two measures with one term having a value of 1
  _____
- Two equivalent ratios or rates
  _____
- Cost of one unit
  _____
- Price of one unit
  _____
- A word meaning *each* or *one*
  _____
- Two or more rates having the same value
  _____
- Two or more ratios having the same value
  _____
- To determine which value is greater or less
  _____

```
I S A L D W C K O H P Z F U M Y
G E E C M L H N B N V Y C T E C
V E E T M R B S G H V Z F U Q Y
Z K N Y A A M U D P O Q Q M C M
E C O R R R X R L G A S H U A X
S O I T A R T N E L A V I U Q E
B M T U L K H N A P E T A R C P
T P R A N V S X E B C C M I G Q
N A O B A I U X Z L N N R A U R
A R P Q K B T B I Y A P L W E A
Q E O O S F J C S A T V M K F A
R D R S W T F V O I O T I R F G
X I P F K U O O N S Z X A U G S
Y U O A K A I U U A T T K N Q F
J U N I T R A T E D I K T W B E
O B U G W I K L I O U T Q N I V
```

| Equivalent ratios | Compare | Unit price | Per | Proportion |
| Unit cost | Rate | Equivalent rates | Unit rate | Ratio |

Standard 6.4(B) – Readiness

Unit 12 Motivation Station

Choices, Choices!

Play *Choices, Choices!* with a partner. Each pair of players needs a game board and a paper clip to use with the spinner. Each player needs a sheet of paper and a pencil. Player 1 spins the spinner and works out the problem spun on his/her paper. Player 1 announces his/her answer and writes it inside the space. If correct, player 1 initials the space along the outside edge. If incorrect, player 1 loses a turn. Player 2 then takes a turn, repeating the process. If the spinner lands on a space that is already initialed, the player moves clockwise to the next open space and solves that problem. The game ends when all spaces have initials. The player who initials more spaces is the winner.

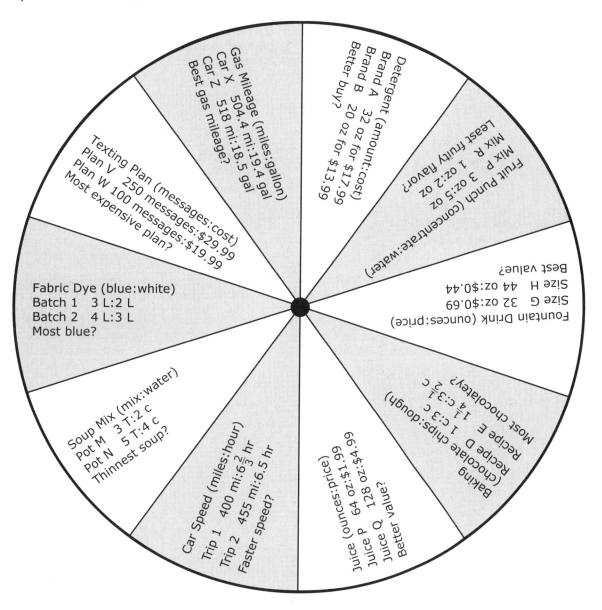

Unit 12 Homework

Standard 6.4(B) – Readiness

Use the information below to answer questions 1–4.

The following combinations of cake mix and water are mixed to make batter.

Mix A – 2 cups cake mix:1 cup water

Mix B – 1 cup cake mix:2 cups water

Mix C – 3 cups cake mix:4 cups water

1. Which batter is the thickest?

   Explain. _____
   _____
   _____

2. Which batter is the thinnest?

   Explain. _____
   _____
   _____

3. If Mix C is the preferred combination, how much cake mix and water is required to make $2\frac{1}{2}$ recipes?

4. Jill starts to make a cake using the ratio in Mix C. She measures her cake mix and only has 2 cups. How much water will she need to add in order to keep the ratio the same?

5. The table shows 4 sizes of bags of potato chips and their prices.

   Potato Chip Prices

Bag	Size	Price
Bag A	10 oz	$1.99
Bag B	15 oz	$2.99
Bag C	18 oz	$3.49
Bag D	22 oz	$4.29

   Bernice purchases Bag D for the class party. Mason purchases Bag C. Which student got the better deal? Why?

   _____
   _____
   _____

   The class needs at least 50 ounces of potato chips for their party. If the teacher decides to purchase all the chips, and purchases all the same size bags, which bag should she purchase?

   _____

   How many bags will she need to purchase?

   Explain your answer.

   _____
   _____

Connections

1. Conduct your own experiment in the kitchen. Test different drinks using a mix or concentrate and water. Keep track of your results in a table until you find the best ratio for a flavorful drink. Share your findings with your class.

2. Help your family become better bargain shoppers. Use a calculator while shopping for groceries or household items to compare prices between different brands and/or different size containers. Write a paragraph about your experience.

Standard 6.4(C) – Supporting

Unit 13 Introduction

Use the picture to answer questions 1–8.

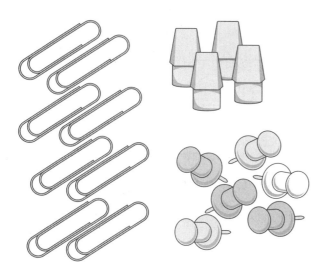

1 What is the ratio of paper clips to erasers?

_____

2 The number of paper clips is _____ times the number of erasers.

3 What is the ratio of erasers to paper clips?

_____

4 The number of erasers is _____ the number of paper clips.

5 What is the ratio of thumbtacks to erasers?

_____

6 The number of thumbtacks is _____ times the number of erasers.

7 What is the ratio of erasers to thumbtacks?

_____

8 The number of erasers is _____ the number of thumbtacks.

Use the table to answer questions 9–12.

The table shows the relationship between the lengths and widths of four rectangles.

Length (in.)	8	12	16	20
Width (in.)	6	9	12	15

9 Write a ratio that describes the multiplicative comparison of width to length.

_____

10 The length of each rectangle is _____ times the width of each rectangle.

11 Based on the table, what is the width of a rectangle with a length of 28 inches?

12 How can the length of a rectangle that has a width of 20 inches be determined?

_____

Use the graph to answer questions 13 and 14.

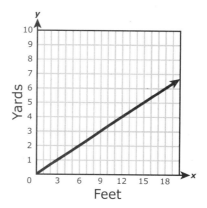

13 The number of yards is _____ times the number of feet.

14 How many yards are equivalent to 20 feet? Show all work.

Unit 13 Guided Practice

Standard 6.4(C) – Supporting

Use the table to answer questions 1 and 2.

The table below shows actual distances compared to distances on a map.

Distances

Actual Distance	Map Distance
14 miles	2 inches
35 miles	5 inches
56 miles	8 inches

1 Which ratio best describes the relationship between the map distance in inches compared to the actual distance in miles?

Ⓐ 7 : 1
Ⓑ 2 : 5
Ⓒ 1 : 7
Ⓓ 5 : 2

2 How many miles does the actual distance measure if the distance on the map measures 14 inches?

Record your answer and fill in the bubbles on the grid below. Be sure to use the correct place value.

3 Mrs. Su is making cookies. The ratio of walnuts to raisins she bakes into each cookie is 3 to 5. If Mrs. Su decides to bake giant cookies, which of the following shows the possible numbers of ingredients in each cookie?

Ⓐ 12 raisins and 20 walnuts
Ⓑ 36 raisins and 60 walnuts
Ⓒ 25 raisins and 15 walnuts
Ⓓ 20 raisins and 15 walnuts

4 Use the drawings below to complete the following statement.

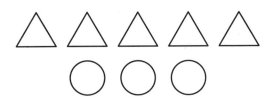

The number of circles is _____ times the number of triangles.

Ⓕ 2
Ⓖ $\frac{3}{5}$
Ⓗ $\frac{2}{3}$
Ⓙ $\frac{5}{3}$

Use the graph to answer questions 5 and 6.

The graph shows the multiplicative comparison for a number of quarts and a corresponding number of gallons.

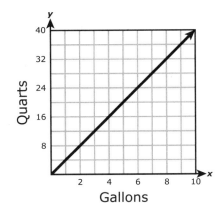

5 Which ratio describes the relationship if the number of gallons is compared to the number of quarts?

Ⓐ 1 : 4
Ⓑ 1 : 1
Ⓒ 4 : 1
Ⓓ 1 : 2

6 How many quarts are equivalent to $14\frac{1}{2}$ gallons?

Ⓕ $3\frac{5}{8}$ quarts
Ⓖ 56 quarts
Ⓗ $46\frac{1}{2}$ quarts
Ⓙ 58 quarts

Standard 6.4(C) – Supporting

Unit 13 Independent Practice

Use the table to answer questions 1 and 2.

The table shows a proportional relationship between x and y.

x	9	15	24	39
y	6	10	16	26

1. Which ratio describes the multiplicative comparison of y to x?

   Ⓐ $\frac{3}{2}$ Ⓒ $\frac{1}{3}$

   Ⓑ $\frac{4}{3}$ Ⓓ $\frac{2}{3}$

2. Find the value of y when x equals 33.

   Record your answer and fill in the bubbles on the grid below. Be sure to use the correct place value.

3. The ratio of boys to girls in the school musical is 5 to 8. Which of the following shows the possible numbers of boys and girls in the school musical?

   Ⓐ 30 girls and 48 boys

   Ⓑ 15 boys and 18 girls

   Ⓒ 20 boys and 32 girls

   Ⓓ Not here

Use the table to answer questions 4 and 5.

The table below shows three different sizes of coffee cups from Starbucks®.

Starbucks® Coffee Cups

Cup	Number of Ounces
Tall	12
Grande	16
Venti	20

4. Which statement is true about the data in the table?

   Ⓕ The tall is $\frac{5}{3}$ the size of the venti.

   Ⓖ The venti is $\frac{4}{5}$ the size of the grande.

   Ⓗ The grande is $\frac{4}{3}$ the size of the tall.

   Ⓙ The venti is $\frac{5}{4}$ times the size of the tall.

5. How many tall cups of coffee will it take to equal the number of ounces in one venti?

   Ⓐ 1.8

   Ⓑ $1\frac{2}{3}$

   Ⓒ $\frac{3}{5}$

   Ⓓ 2

Unit 13 Assessment

Name _____

Standard 6.4(C) – Supporting

1 The ratio of white socks to black socks in Terry's drawer is $\frac{4}{5}$. Which of the following combinations could NOT be the actual numbers of white and black socks in Terry's drawer?

Ⓐ 25 black socks; 20 white socks

Ⓑ 35 black socks; 28 white socks

Ⓒ 24 black socks; 30 white socks

Ⓓ 15 black socks; 12 white socks

Use the table to answer questions 2 and 3.

Ashley's grandfather encourages her to save her nickels. He matches the nickels she saves with a number of dimes.

Coins

Nickels	Dimes
10	15
24	36
40	60

2 Which ratio describes the multiplicative relationship of dimes to nickels?

Ⓕ 2 to 3

Ⓖ $\frac{5}{12}$

Ⓗ 4 to 3

Ⓙ 3:2

3 If Ashley receives 150 dimes from her grandfather, how many nickels did she save?

Ⓐ 50

Ⓑ 100

Ⓒ 175

Ⓓ 225

4 Use the integers below to complete the following statement.

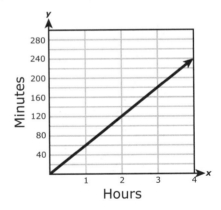

The number of negative integers is _____ times the number of positive integers.

Ⓕ $\frac{3}{4}$ Ⓗ $\frac{4}{3}$

Ⓖ $\frac{2}{3}$ Ⓙ $\frac{3}{2}$

5 The graph below shows the relationship between hours and minutes.

What is the multiplicative relationship shown in the graph?

Ⓐ One minute is 60 times one hour because $\frac{3}{180} = 60$.

Ⓑ One minute is $\frac{1}{60}$ times one hour because $\frac{180}{3} = \frac{1}{60}$.

Ⓒ One hour is 60 times one minute because $\frac{180}{3} = 60$.

Ⓓ One hour is $\frac{1}{60}$ times one minute because $\frac{3}{180} = \frac{1}{60}$.

Standard 6.4(C) – Supporting

Unit 13 Critical Thinking

1. Margaret mixes red and blue paint to make purple. She finds the best ratio to get the shade of purple she wants is 2 quarts of red paint and 3 quarts of blue paint. Margaret needs a total of 45 quarts of purple paint. She begins a table to help organize the data. Margaret is not sure what to do next to find the number of quarts of red and blue paint she needs.

Quarts of Red Paint	Quarts of Blue Paint	Quarts of Purple Paint
2	3	5
4	6	10
		20
		45

What part of the purple paint is red paint? _____

The amount of purple paint is _____ times the amount of red paint.

What part of the purple paint is blue paint? _____

The amount of purple paint is _____ times the amount of blue paint.

Explain to Margaret how to use your answers above to complete the table.

_____

_____

Complete the table.

How many quarts of red paint does Margaret need to make 45 quarts of purple paint? _____

How many quarts of blue paint does Margaret need to make 45 quarts of purple paint? _____

2. Use the ratio 3:4 to write an original word problem. You may use a table, graph, or drawing to support your problem. Include four answer choices with your problem.

Unit 13 Journal/Vocabulary Activity

Standard 6.4(C) – Supporting

Journal

Explain what the ratio 3 cats to 2 dogs means. Use the vocabulary words *ratio*, *multiply* or *multiplication*, and *comparison* in the explanation.

Vocabulary Activity

For each of the boxes, fill in the blanks to show the *multiplicative comparison* between the two *quantities*. Then fill in the correct *ratios*.

There are _____ or _____ times as many girls as boys.

There are _____ times as many boys as girls.

The ratio of girls to boys is _____.

The ratio of boys to girls is _____.

There are _____ times as many bananas as pears.

There are _____ times as many pears as bananas.

The ratio of bananas to pears is _____.

The ratio of pears to bananas is _____.

There are _____ or _____ times as many books as book bags.

There are _____ times as many book bags as books.

The ratio of books to book bags is _____.

The ratio of book bags to books is _____.

There are _____ times as many socks as shoes.

There are _____ times as many shoes as socks.

The ratio of socks to shoes is _____. The ratio of shoes to socks is _____.

Standard 6.4(C) – Supporting

Unit 13 Motivation Station

Color the Number

For each section in the picture below, a ratio is given. Color the sections according to the rules given at the bottom of the page.

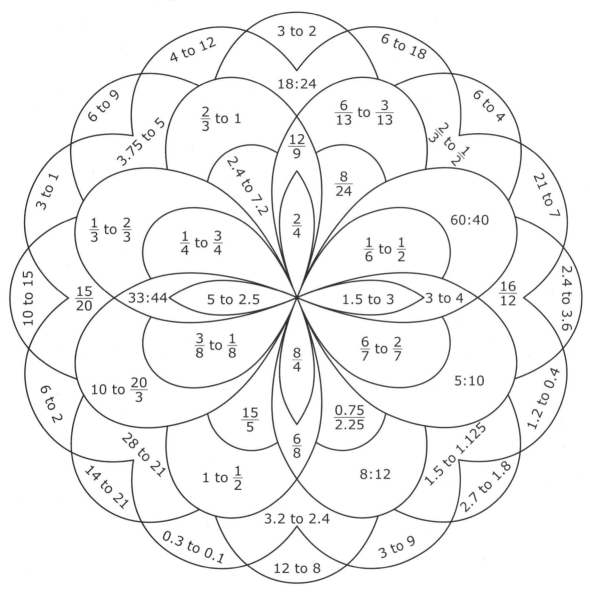

If the relationship shows the first value is $\frac{1}{2}$ the second value OR 2 times the second value, color the section red.

If the relationship shows the first value is $\frac{2}{3}$ the second value OR $\frac{3}{2}$ times the second value, color the section blue.

If the relationship shows the first value is $\frac{1}{3}$ the second value OR 3 times the second value, color the section green.

If the relationship shows the first value is $\frac{3}{4}$ the second value OR $\frac{4}{3}$ times the second value, color the section purple.

Unit 13 Homework

Standard 6.4(C) – Supporting

1. A piano has black and white keys like those shown in the picture below.

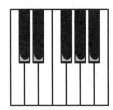

Using the keys in the picture, complete the following statements.

There are _____ times as many black keys as white keys.

There are _____ times as many white keys as black keys.

The ratio of black keys to white keys is _____.

2. Marco and Lydia both have cookies as shown.

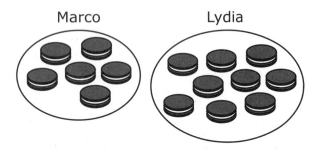

How many stacks of Marco's cookies will it take to equal one stack of Lydia's cookies?

_____

3. There are 1.5 times as many girls as boys in Mr. Cooper's class. If there are 12 boys in Mr. Cooper's class, how many girls are in the class? Show all work.

4. Ella has the following items in her bag.

Complete the following statements about the items in Ella's bag.

There are _____ times as many CDs as water bottles in Ella's bag.

There are _____ times as many pens as apples in Ella's bag.

There are _____ times as many pens as CDs in Ella's bag.

There are _____ times as many water bottles as apples in Ella's bag.

There are _____ times as many apples as pens in Ella's bag.

There are _____ times as many CDs as apples in Ella's bag.

There are _____ times as many water bottles as pens in Ella's bag.

There are _____ times as many apples as water bottles in Ella's bag.

Connections

1. Find ratios around your house. Some examples might include the ratio of pets to people, the ratio of people to rooms, or the ratio of doors to windows in your house. Record four ratios and then write two different multiplicative comparisons for each.

2. Work with a friend or family member to brainstorm ways ratios are used in the real world. Select three different examples to share with the class.

Name _____

Standard 6.4(D) – Supporting

Unit 14 Introduction

1. The Cupcakery sells a half-dozen specialty cupcakes for $16.50. Write a rate to determine the cost per cupcake.

   _____

   What is the cost per cupcake? _____

2. The table below shows the distances and times it took 4 bus drivers to transport children to Camp Sunshine from 4 different cities.

   Travel to Camp Sunshine

Driver	Time	Distance
Mr. Sanchez	4 hours	188 miles
Mrs. Mills	3 hours	168 miles
Mr. Holloway	5 hours	315 miles
Mrs. Alexander	2 hours	112 miles

   What does the expression 188 ÷ 4 represent?

   _____
   _____

   Explain how you know.

   _____
   _____

   Write an expression to find the distance Mrs. Mills travels in one hour.

   _____

   Which driver's rate is 63 miles per hour? Show all work.

3. John can read 135 pages in $1\frac{1}{2}$ hours. Write John's reading rate in pages per minute.

   _____
   _____

4. Cynthia purchased 3 video games from the sale bin that were each the same price. She also bought 2 DVDs for $15 each. Cynthia spent a total of $117. Write the first step needed to find the unit price for each video game.

   _____

   Write and solve an expression to find the price for each video game.

   _____

5. Justin played Space Invaders at the video arcade. He played for 45 minutes and scored 4,815 points. What was the average number of points Justin scored during each minute of the game?

   _____

6. Ahrom spent $9 at the gas station for 12 candy bars. Each candy bar has the same price. Explain the meaning of each rate below using information from the problem.

   $\frac{9}{12}$ _____

   $\frac{12}{9}$ _____

Unit 14 Guided Practice

Standard 6.4(D) — Supporting

1 Jessie buys a case of drinks for $5.98. A case of drinks contains 24 eight-ounce cans. Which expression could Jessie use to determine the price he paid for each eight-ounce can?

Ⓐ $\frac{5.98}{24}$ Ⓒ $\frac{24}{5.98}$

Ⓑ $\frac{5.98}{8}$ Ⓓ $\frac{8}{5.98}$

2 Braden's parents pay $780 per year for him to participate in karate classes. Braden has one karate class each week. Which of the following shows the cost per karate class?

Ⓕ $2.14, because there are 365 days in a year

Ⓖ $65, because $\frac{780}{12} = 65$

Ⓗ $52, because there are 52 weeks in a year

Ⓙ $15, because $\frac{780}{52} = 15$

Use the table to answer questions 3 and 4.

The table shows the cost Chet paid for different numbers of song downloads from two different music websites.

Song Downloads

Music Sites	Number of Songs	Cost
Get Tunes	27	$13.50
Music-N-More	16	$12.00

3 How many songs can Chet download for $1 using Get Tunes?

Ⓐ $\frac{1}{2}$ song Ⓒ 2 songs

Ⓑ 1 song Ⓓ 3 songs

4 What is the rate per song for downloading music from Music-N-More?

Ⓕ $0.50 Ⓗ $1.33

Ⓖ $0.75 Ⓙ $1.25

Use the table to answer questions 5 and 6.

Four friends record the number of pages and amount of time they spend reading. The results are displayed below.

Reading Results

Name	Pages Read	Number of Hours
Ashley	648	4.5
Simon	455	3.5
Greyson	246	1.5
Jessica	400	2.5

5 Which rate results in the most number of pages read per hour?

Ⓐ 4.5 ÷ 648 Ⓒ 3.5 ÷ 455

Ⓑ 400 ÷ 2.5 Ⓓ 246 ÷ 1.5

6 Which of the following statements is NOT true?

Ⓕ The expression $\frac{270}{648}$ can be used to find the number of minutes it takes Ashley to read one page.

Ⓖ The expression $\frac{455}{3.5}$ can be used to find Simon's reading rate per hour.

Ⓗ The expression $\frac{246}{1.5}$ can be used to find Greyson's reading rate per minute.

Ⓙ The expression $\frac{2.5}{400}$ can be used to find the number of hours it takes Jessica to read one page.

7 Rosa and her 7 friends equally share two pizzas. If each pizza has 12 slices, which best describes the amount of pizza each received?

Ⓐ 3.5 slices per person

Ⓑ 3 slices per person

Ⓒ 1.7 slices per person

Ⓓ 4 slices per person

Standard 6.4(D) – Supporting

Unit 14 Independent Practice

1 Travis lives 195 miles from Dallas. On a normal trip, it takes 180 minutes to drive from his house to Dallas. What is Travis' average rate of speed in miles per hour?

Ⓐ 65 miles per hour

Ⓑ 1.08 miles per hour

Ⓒ 75 miles per hour

Ⓓ 15 miles per hour

2 The chart below shows the prices of four different packages of socks.

Sock Prices

Package	Pairs per Package	Price
A	5	$4.00
B	8	$6.00
C	9	$9.00
D	10	$7.00

Which of the following statements is true?

Ⓕ The price per pair of socks in package A is $1.25.

Ⓖ Each pair of socks in package B costs $0.75.

Ⓗ Package C is the best buy.

Ⓙ The socks in package D cost $1.43 per pair.

Use the information to answer questions 3 and 4.

Mrs. Calloway used the recipe shown below to make peanut butter cookies.

> Peanut Butter Cookies
>
> $\frac{3}{4}$ cup peanut butter
>
> $\frac{2}{3}$ cup sugar
>
> 1 egg
>
> $\frac{1}{4}$ tablespoon vanilla extract

3 If the recipe makes $1\frac{1}{2}$ dozen cookies, which expression could be used to determine the cups of sugar needed per cookie?

Ⓐ $1\frac{1}{2} \div \frac{2}{3}$ Ⓒ $\frac{2}{3} \div 1\frac{1}{2}$

Ⓑ $18 \div \frac{2}{3}$ Ⓓ $\frac{2}{3} \div 18$

4 What does the rate shown below represent?

$$\frac{3}{4} \div \frac{3}{2}$$

Ⓕ The number of dozens of cookies for each cup of peanut butter.

Ⓖ The number of cookies for each cup of peanut butter.

Ⓗ The amount of peanut butter in one dozen cookies.

Ⓙ The amount of peanut butter in each cookie.

5 Mari baked 4 dozen tarts in 3 hours. At this rate, how many tarts did Mari bake per hour?

Ⓐ $1\frac{1}{3}$, because $\frac{4}{3} = 1\frac{1}{3}$

Ⓑ $\frac{1}{16}$, because $\frac{3}{48} = \frac{1}{16}$

Ⓒ 16, because $\frac{48}{3} = 16$

Ⓓ 12, because 4(3) = 12

Unit 14 Assessment

Standard 6.4(D) – Supporting

1 Rosie's Rose-a-Rama is offering a special price on roses for Valentine's Day. The table below shows the prices for four different rose arrangements.

Valentine Rose Arrangements

Number of Roses	Price of Arrangement
1 dozen	$33.00
15	$30.00
18	$45.00
2 dozen	$54.00

Which statement is true?

Ⓐ The roses in the arrangement with 15 roses cost $0.50 each.

Ⓑ The cost per rose in the 2 dozen arrangement is $2.25.

Ⓒ Each rose costs $0.40 if 18 roses are purchased.

Ⓓ One dozen roses is the best buy.

2 Mr. Aguilar purchases gas for $3.24 per gallon. After leaving the gas station, he travels 208 miles on 8 gallons of gas. If the car uses gas at a constant rate, which best describes the number of miles traveled per gallon of gas?

Ⓕ 84 mpg, because 208 • 3.24 ÷ 8 = 84

Ⓖ 8 mpg, because 208 ÷ (3.24 • 8) = 8

Ⓗ 26 mpg, because 208 ÷ 8 = 26

Ⓙ 64 mpg, because 208 ÷ 3.24 = 64

3 Brittany's scout troop is selling cookies. The scout leader orders 12 cases of cookies, with 12 boxes of cookies per case. There are 16 girls in the troop, and each girl receives an equal number of boxes. Which expression could be used to determine the number of boxes of cookies Brittany receives?

Ⓐ $\frac{12}{16}$ Ⓒ $\frac{16}{12}$

Ⓑ $\frac{12 \cdot 12}{16}$ Ⓓ $\frac{16}{12 \cdot 12}$

4 Kaliah's Hair Salon has shampoo on sale this week as shown in the table below.

Shampoo Sale

Bottle Size	Cost
8 ounces	$5.60
16 ounces	$8.00
24 ounces	$11.52
48 ounces	$28.80

Which size bottle offers shampoo at $0.48 per ounce?

Ⓕ 8-ounce bottle   Ⓗ 24-ounce bottle

Ⓖ 16-ounce bottle   Ⓙ 48-ounce bottle

5 Helen bought ice cream for her friends after school. She bought 4 single-scoop ice cream cones for a total of $8.25 and 6 sundaes. Each sundae cost the same amount, and Helen paid a total of $33.75 for the ice cream. What was the price per sundae Helen paid?

Record your answer and fill in the bubbles on the grid below. Be sure to use the correct place value.

Standard 6.4(D) – Supporting

Unit 14 Critical Thinking

1. Lourdes sews costumes for the school play. It takes her approximately 8 hours to sew the first 3 identical costumes. How long does it take Lourdes to sew 1 costume? Give your answer in hours and minutes.

_____

Does Lourdes complete an entire costume in 1 hour? Explain your answer. _____

_____

_____

Which unit rate would be most useful, the amount of time per costume or the number of costumes per hour? _____

Why? _____

_____

2. At the farmers' market, Farmer Brown sells tomatoes at 5 pounds for $7. Farmer Jones sells his tomatoes at 4 pounds for $6.

   Explain how rates are used to determine which farmer has the better price on tomatoes.

_____

_____

_____

Joanna finds that Farmer Jones has the better price on tomatoes, $0.67 per pound. Is Joanna correct? Show your work below to justify your answer.

Explain why it is important to know which number is the dividend and which is the divisor when finding a rate or unit rate.

_____

_____

_____

Unit 14 Journal/Vocabulary Activity

Name _____

Standard 6.4(D) – Supporting

Journal

Explain how a rate is different from a ratio.
_____
_____

How does knowing how to find a unit rate benefit you?
_____
_____
_____
_____

Vocabulary Activity

Fill in the blanks below.

1. A unit rate is found using _____. When written in fraction form, the _____ is the dividend, and the _____ is the divisor. The _____ is the unit rate.

Label each example with the terms *dividend, divisor,* and *quotient*.

2. $a \div b = c$ a = _____   b = _____   c = _____

3. $\frac{c}{a} = b$ a = _____   b = _____   c = _____

4. $c\overline{)b}^{\,a}$ a = _____   b = _____   c = _____

For each of the following, write whether the situation best describes a *rate* or a *unit rate*.

5. $6.99 per pound _____
6. $3.99 for 5 cans _____
7. 12 horses in 2 pastures _____
8. 52 students in 4 classes _____
9. $0.99 for 1 can _____

10. 32 kids on each bus _____
11. 650 miles per day _____
12. 48 laps in 6 days _____
13. $3.37 per gallon _____
14. 18 feet in 6 minutes _____

Standard 6.4(D) – Supporting

Unit 14 Motivation Station

Spin the Rate

Play *Spin the Rate* with a partner. Each pair of players needs a number cube, a playing board, two different tokens, and a paper clip to use with the spinners. Each player needs a sheet of paper and a pencil. Player 1 spins the *Quantity* spinner and rolls the number cube. On his/her paper, player 1 records a rate using the number spun as the numerator and the number rolled as the denominator. Then player 1 spins the *Labels* spinner. Player 1 uses the label spun to correctly label the rate written on the paper. For example, if a player spins 8, rolls a 4, and spins *Dollars/Pound*, he/she would write "$8 for 4 pounds" on the paper. If the player has correctly written the rate with the labels, he/she moves the token one space on the game board. If the rate is a unit rate, the player advances two additional spaces. If the player incorrectly writes the rate, he/she loses a turn and play passes to player 2. The first player to reach *End* wins the game.

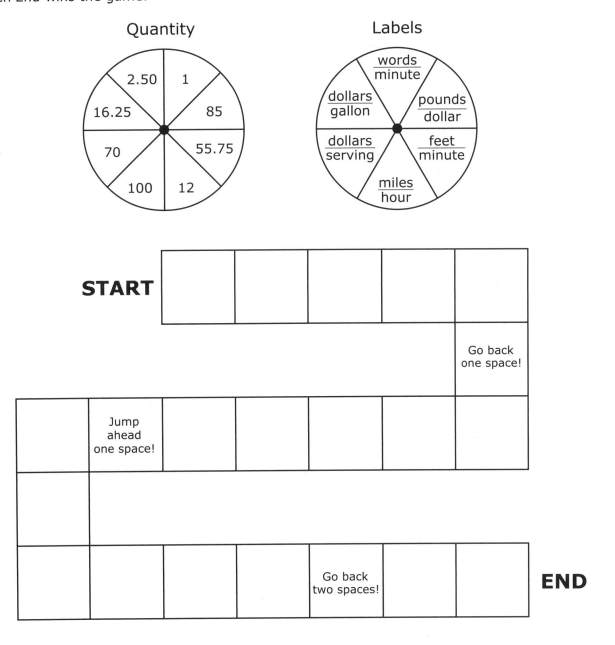

Unit 14 Homework

Standard 6.4(D) – Supporting

1. Nicholas and Brenda work on their math homework together. For one problem, they are asked to find the number of miles a car travels in 1 hour. They know that the car travels 330 miles in $5\frac{1}{2}$ hours. Nicholas divides 330 by $5\frac{1}{2}$ to get the rate. Brenda disagrees with Nicholas and divides $5\frac{1}{2}$ by 330 to find the rate. Which student is correct? Explain.

   _____

   _____

2. Five gallons of gas costs $16.25. What is the cost per gallon of the gas?

   What unit rate describes the number of gallons per dollar?

   How are the two rates different?

   _____

   _____

   Why might it be important to know both of these rates in real life?

   _____

   _____

Use the label to answer questions 3 and 4. Express answers as fractions in simplest form.

Nutrition Facts		
Serving Size 1 ounce (28 g, approx. 16 crisps)		
Servings Per Container	approx. 5	
Amount Per Serving		
Calories	150	
Calories from Fat	80	
		% Daily Value
Total Fat	9 g	14%
Saturated Fat	2.5 g	13%
Trans Fat	0 g	
Cholesterol	0 mg	0%
Sodium	190 mg	8%
Total Carbohydrate	15 g	5%
Dietary Fiber	1 g	4%
Sugars	1 g	

3. According to the information provided in the label, what is the rate of calories per crisp?

   _____

   What is the rate of crisps per calorie?

   _____

   Which rate is most useful if a person has only 300 calories left on their diet and wants a snack? Explain.

   _____

   _____

4. Calculate each of the following:

   • number of calories per fat gram

   • number of calories per gram of carbohydrates

   • number of grams of saturated fat per gram of fat

Connections

1. On the next shopping trip with your family, pick out your two favorite cereals. What is the cost per ounce for the smallest box of each cereal? What is the cost per ounce for the largest box of each cereal? How does knowing this information help you make a decision about which cereal and size to purchase?

2. Look through the weekly ad from your local grocery store. Cut out all of the ads that involve rates. Use the ads to make a unit rate poster. Find the unit rate for each ad and write it under the picture. Share your poster with the class.

Name _____

Standards 6.4(E) – Supporting, 6.4(F) – Supporting

Unit 15 Introduction

1. Look at the model below.

What percent of the model is shaded?
_____

What fraction of the model is NOT shaded?
_____

2. Mrs. Cupit surveyed her class to determine how many students ride the bus. The survey results showed that 9 out of 24 students ride the bus to school. Represented as a decimal, what part of Mrs. Cupit's class rides the bus?

_____

Explain how you found the answer.

_____

_____

3. Using the number line below, what percent of the distance from 0 to 1 is represented by Point C?

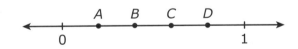

_____

4. Reginald answered 78% of the questions on his social studies quiz correctly. What fractional part of the questions did Reginald NOT answer correctly?

_____

5. Shade the model below to represent $\frac{1}{4}$.

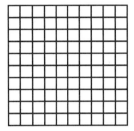

What percent of the model did you shade?

_____

Explain your answer.

_____

6. Mitchell sold 56% of his baseball cards to buy a new bike. What decimal represents the part of Mitchell's baseball cards he sold?

_____

7. Shade each model to represent the number shown.

a. $\frac{4}{5}$

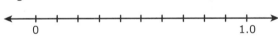

b. 0.2

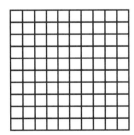

c. 62.5%

Unit 15 Guided Practice

Standards 6.4(E) – Supporting, 6.4(F) – Supporting

1. Roderick correctly answers 92% of the questions on his math benchmark test. Which fraction represents the part of questions Roderick answered correctly?

   Ⓐ $\frac{22}{25}$ Ⓒ $\frac{23}{25}$

   Ⓑ $\frac{19}{20}$ Ⓓ $\frac{18}{20}$

2. Each model below is divided into sections of equal size. Which model does NOT have 75% of the total area shaded?

   Ⓕ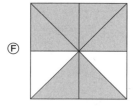

   Ⓖ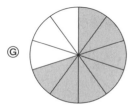

   Ⓗ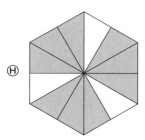

   Ⓙ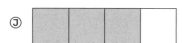

3. Shelby earned $20 babysitting. She spent 30% of her earnings at the mall. What decimal represents the part of her earnings Shelby spent?

   Ⓐ 0.3   Ⓒ 0.03

   Ⓑ $0.\overline{3}$ Ⓓ 30.0

4. The shaded region of the model represents 10%.

   Which model best represents a shaded region of 40%?

   Ⓕ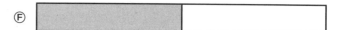

   Ⓖ

   Ⓗ

   Ⓙ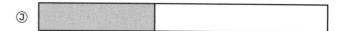

5. The sixth grade at Carter Middle School conducted a survey of 300 students to determine their favorite ice cream flavor. Of the students surveyed, 20% selected chocolate chip cookie dough. Which number line best represents the part of the student body that selected chocolate chip cookie dough as their favorite?

   Ⓐ

   Ⓑ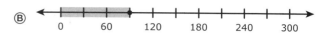

   Ⓒ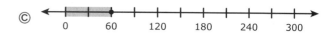

   Ⓓ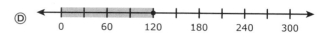

Name _____

Standards 6.4(E) – Supporting, 6.4(F) – Supporting

Unit 15 Independent Practice

1 The shaded region of the model below represents the part of the students who won ribbons for competing in the science fair.

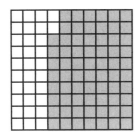

What percent of the students who competed in the science fair did NOT win ribbons?

Ⓐ 68%   Ⓒ 34%

Ⓑ 42%   Ⓓ 32%

2 After repairing several water leaks in their house, the Foster family noticed a 10% decrease in their monthly water bill. What decimal number represents the percent decrease in the Foster family's water bill?

Record your answer and fill in the bubbles on the grid below. Be sure to use the correct place value.

3 If the shaded region of the model below represents the part of a book that Katie has read, what fraction of the book does Katie have left to read?

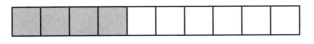

Ⓐ $\frac{3}{10}$ Ⓒ $\frac{3}{5}$

Ⓑ $\frac{4}{5}$ Ⓓ $\frac{7}{10}$

4 Which of the following represents equivalent values?

Ⓕ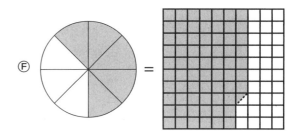

Ⓖ 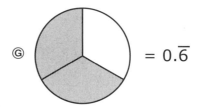 = $0.\overline{6}$

Ⓗ 5% =

Ⓙ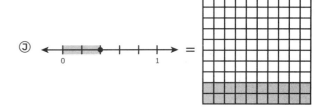

5 A farmer has 50 acres of corn crops. He harvested 3 acres of his crop in one afternoon. Which decimal represents the part of the corn crop the farmer harvested?

Ⓐ 0.06

Ⓑ 3.50

Ⓒ 0.60

Ⓓ 3.00

Unit 15 Assessment

Name _____

Standards 6.4(E) – Supporting, 6.4(F) – Supporting

1. Which of the following represents 40%?

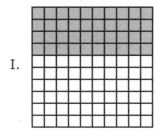
I.

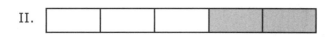
II.

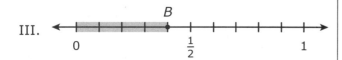
III.

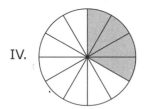
IV.

Ⓐ I only

Ⓑ I and II only

Ⓒ I, II, and III only

Ⓓ I, II, III, and IV

2. James divides a piece of poster board into equal sections and uses the shaded sections for an art project.

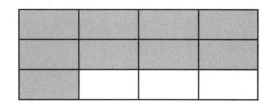

What percent of the poster board does James have left for other projects?

Ⓕ 75%   Ⓗ 25%

Ⓖ 9%   Ⓙ 3%

3. Gretchen completes 36 out of 100 integer homework problems during class. Which fraction represents the part of Gretchen's completed homework assignment?

Ⓐ $\frac{6}{25}$ Ⓒ $\frac{27}{50}$

Ⓑ $\frac{3}{8}$ Ⓓ $\frac{9}{25}$

4. Lynn shaded the figure below.

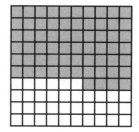

How would this amount be represented as a ratio and a percent?

Ⓕ 64 out of 100, 6.4%

Ⓖ 16 out of 25, 64%

Ⓗ 64 out of 100, 640%

Ⓙ 19 out of 25, 64%

5. Each square below is divided into sections of equal size. Which square has 37.5% of its total area shaded?

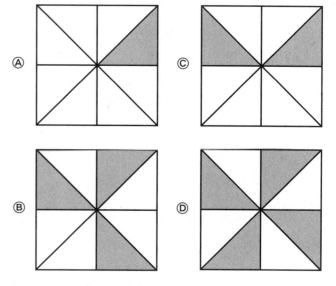

Name _____

Standards 6.4(E) – Supporting, 6.4(F) – Supporting

Unit 15 Critical Thinking

1 In Mr. Meyer's class, $\frac{2}{3}$ of the students are on the honor roll. There are 10 students not on the honor roll. Draw a strip diagram below to model the part of the class that is on the honor roll.

How many students are on the honor roll? _____

Explain how you used the diagram to find the number of students on the honor roll.

_____

_____

2 Liz set a goal to sell 100 candy bars for the spring band trip. After one week, she sold 120 candy bars. Based on her goal of 100 candy bars, shade 10 × 10 grids to represent the percent of the goal Liz met.

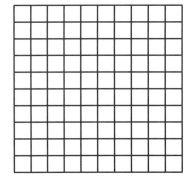
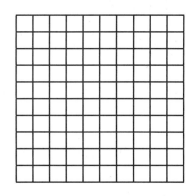
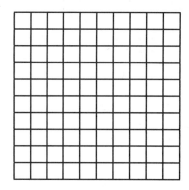

How many grids did you use to represent Liz's goal? _____ Why? _____

_____

Would it ever be necessary to shade less than one small square on a 10 × 10 grid to represent a percent? _____ Why? Give an example from everyday life.

_____

_____

Unit 15 Journal/Vocabulary Activity

Name _____

Standards 6.4(E) – Supporting, 6.4(F) – Supporting

Journal

Explain to a fifth-grade student how to use a strip diagram to represent a fraction or a percent.

_____
_____
_____

Vocabulary Activity

For this activity, work with a partner. Each pair of players needs an activity page and a paper clip to use with the spinner. Each player needs a sheet of paper and a pencil. Player 1 secretly writes a vocabulary term from the list on the paper, and then spins the spinner and follows the directions for the secret term. Player 2 tries to guess the vocabulary term. If the term is guessed correctly, it is marked off the list. Players switch roles and continue until all vocabulary terms have been marked off the list.

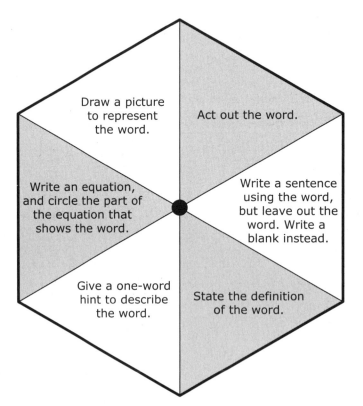

Vocabulary Terms

ratio	fraction	percent	fraction model
mixed number	simplest form	denominator	tenths
percent model	numerator	benchmark	hundredths

Name _____

Standards 6.4(E) – Supporting, 6.4(F) – Supporting

Unit 15 Motivation Station

Tic-Tac-Representation

Play *Tic-Tac-Representation* with a partner. Each pair of players needs a game board and a paper clip to use with the spinners. Each player needs a pencil, a sheet of paper, and several tokens (a different color for each student). Player 1 spins each of the spinners and writes the fraction or percent and the representation on his/her paper. Player 1 then creates the appropriate representation on the paper. If the drawing is correct, player 1 chooses a space on the game board with the same representation as the one created and places a token on it. If the drawing is incorrect, player 1 loses a turn. Player 2 repeats the process. The first player to cover 4 spaces in a row horizontally, vertically, or diagonally wins the game.

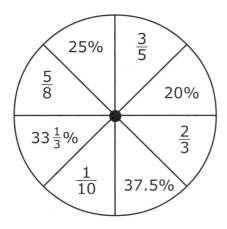 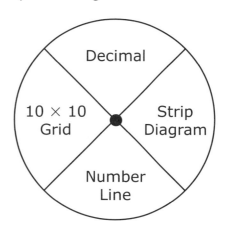

10 × 10 Grid	Decimal	Strip Diagram	Number Line
Strip Diagram	Number Line	10 × 10 Grid	Decimal
Decimal	10 × 10 Grid	Number Line	Strip Diagram
Number Line	Strip Diagram	Decimal	10 × 10 Grid

©2014 mentoringminds.com   motivationmath™ LEVEL 6   ILLEGAL TO COPY   125

Unit 15 Homework

Standards 6.4(E) – Supporting, 6.4(F) – Supporting

1. Lorienne read 60 pages of her library book over the weekend. The book has a total of 180 pages. Draw a model below to represent the part of the library book Lorienne read over the weekend.

2. Manuel has completed 62.5% of his work towards becoming a certified scuba diver. Shade and label the strip diagram below to represent the remaining amount of work Manuel has to complete for his certification.

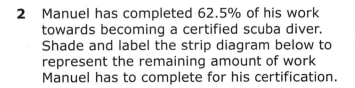

0%   100%

3. Beau took 9 of the 12 eggs in a carton and boiled them to make deviled eggs. Expressed as a decimal, what part of the carton of eggs did Beau boil?

   _____

   Explain how you found your answer.

   _____
   _____

4. Write three fractions that are equivalent to 30.2% using denominators of 1,000, 100, and 10.

   _____
   _____

5. For each of the following, complete the representation.

   a. $33\frac{1}{3}\%$

   0%   100%

   b. $\frac{3}{4}$

   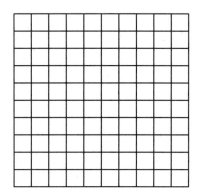

   c. 60%

   ```
   ←——————+————————————————+——→
          0                1
   ```

   d. $\frac{7}{8}$

   [empty bar]

   0   1

Connections

1. Think about where you encounter ratios and percents in your daily life. When you are helping in the kitchen, working on a construction or automotive project, or shopping, look for examples of ratios and percents. Think about how you would represent these as fractions or decimals or by using a model. Share your experiences with the class.

2. Find three examples in a science textbook that use a percent less than 1. On a sheet of paper, write each percent and a brief description of what the number is about. Then draw a 10 × 10 grid model for each percent. Share your findings with the class.

Standard 6.4(G) – Readiness

Unit 16 Introduction

1. Gabe put 100% of the money he earned mowing lawns into a savings account. Write a decimal number that is equivalent to 100%.

   _____

2. Three-fourths of the 28 students in Mrs. Mobley's class brought their lunches to school today. What percent is equivalent to the part of Mrs. Mobley's students that brought a lunch?

   _____

3. In the last school board election, Mrs. Jackson received 60.1% of the votes. Write three equivalent fractions for 60.1% with denominators of 1,000, 100, and 10.

   _____

4. Megan paints her locker red, white, and blue. She paints $\frac{9}{20}$ of the locker red, 15% of the locker white, and 0.4 of the locker blue. Complete the table below.

	Fraction	Decimal	Percent
Red	$\frac{9}{20}$		
White			15%
Blue		0.4	

5. DeMarco has the following coins in his pocket: 5 nickels, 3 dimes, and 2 quarters. What percent of one dollar does DeMarco have in nickels?

   _____

Use the table to answer questions 6–9.

Coach Spitzer surveys 50 students in his class to determine which new sport they want to add to the athletic schedule next year. The table shows the results of the survey.

New Sports

Sport	Part of Votes
Soccer	28%
Rugby	6 students
Swimming	18%
Tennis	0.24
Baseball	$\frac{9}{50}$

6. Explain how to determine the percent of students who voted for rugby as the new sport.

   _____
   _____

   What percent of students voted for rugby?

   _____

7. Which sports received an equivalent amount of votes?

   _____

8. What fraction of the students want tennis as the new sport?

   _____

9. What decimal represents the part of students who voted for soccer?

   _____

Unit 16 Guided Practice

Standard 6.4(G) – Readiness

1. Tameka has 4 coins in her purse. The value of the coins is 30% of a dollar. Which could represent the coins in Tameka's purse?

   Ⓐ 4 nickels

   Ⓑ 2 dimes and 2 nickels

   Ⓒ 1 quarter, 1 dime, and 2 nickels

   Ⓓ 3 dimes and 1 nickel

2. Mark, Rea, Holly, and Travis compete in a contest to see who can bike around the school the fastest. Mark's finishing time is 3.125 minutes, Rea's time is $3\frac{8}{60}$ minutes, Holly's time is $3\frac{1}{8}$ minutes, and Travis' time is $\frac{24}{7}$ minutes. Which students complete the contest in the same amount of time?

   Ⓕ Mark and Rea   Ⓗ Travis and Rea

   Ⓖ Holly and Travis   Ⓙ Mark and Holly

3. Jolie surveys the students in her class to determine their favorite frozen custard flavor. The table shows the results.

   Custard Flavors

Flavor	Number of Students
Vanilla	8
Chocolate	10
Strawberry	5
Lemon	2

   Which statement is true?

   Ⓐ The part of students who like vanilla is represented by the equation $\frac{8}{25} = 0.34$.

   Ⓑ The part of students who like chocolate is represented by the equation $\frac{2}{5} = 20\%$.

   Ⓒ The part of students who like strawberry is represented by the equation $\frac{1}{5} = 0.05$.

   Ⓓ The part of students who like lemon is represented by the equation $\frac{2}{25} = 8\%$.

4. Fernando fills a one-fourth cup measuring tool with sugar 5 times and pours it into a punch bowl. Which represents the sugar Fernando adds to the punch bowl?

   Ⓕ 1.5 c   Ⓗ 1.25 c

   Ⓖ $\frac{1}{4}$ c   Ⓙ $\frac{5}{20}$ c

5. During Centerville Middle School's annual physical fitness tests, the physical education teacher found that $62\frac{1}{2}\%$ of the students exceeded the minimum physical fitness standards. Which represents the part of the students who exceeded the standards?

   Ⓐ 0.0625   Ⓒ 62.5

   Ⓑ $\frac{625}{1000}$ Ⓓ $\frac{7}{9}$

6. Malik completes 0.7 of the total number of push-ups his coach requires. Which does NOT represent the part of the push-ups Malik completes?

   Ⓕ 7%   Ⓗ 70%

   Ⓖ $\frac{14}{20}$ Ⓙ 0.70

7. Mrs. Martin wrote three equations on the board.

   - Equation 1: $3\frac{4}{5} = 3.8$
   - Equation 2: $24\% = \frac{6}{25}$
   - Equation 3: $12.5 = 125\%$

   Which of the equations is NOT true?

   Ⓐ Equation 1 only

   Ⓑ Equations 2 and 3 only

   Ⓒ Equation 3 only

   Ⓓ All three equations are true.

Name _____

Standard 6.4(G) – Readiness

Unit 16 Independent Practice

1. Mr. Gazette plants a tulip bulb that has a diameter of $\frac{1}{25}$ meter. Which shows an equivalent length?

   Ⓐ $\frac{2}{75}$ m
   Ⓑ 0.04 m
   Ⓒ $1\frac{2}{5}$ m
   Ⓓ 0.125 m

2. Ramon correctly answered 80% of the questions on his math test. Which of the following does NOT represent the part of the questions Ramon answered correctly?

   Ⓕ $\frac{12}{15}$
   Ⓖ 0.80
   Ⓗ $\frac{28}{35}$
   Ⓙ $\frac{34}{40}$

3. Mrs. Lin surveys her class to find their favorite type of cookie. The results are shown in the table below.

   Cookie Survey

Type of Cookie	Number of Students
Chocolate chip	8
Peanut butter	5
Oatmeal	3
Sugar	4

   Which shows the part of students whose favorite cookie is peanut butter?

   Ⓐ $\frac{1}{5}$
   Ⓑ 0.25
   Ⓒ $\frac{1}{3}$
   Ⓓ 0.20

4. Mrs. McMichael paid for her groceries with cash. The change she received from the cashier was 135% of a dollar. Which best represents the possible change Mrs. McMichael received?

   Ⓕ 5 quarters and 1 nickel
   Ⓖ 1 half-dollar, 3 quarters, 1 dime, and 1 nickel
   Ⓗ 4 quarters, 3 dimes, and 1 nickel
   Ⓙ 1 dollar, 2 dimes, and 2 nickels

5. Kendall opens a savings account at First Federal Savings. The account earns interest at a rate of $\frac{1}{2}$% annually. Which of the following is equivalent to the interest rate Kendall earns on her savings account?

   Ⓐ 0.5
   Ⓑ $\frac{1}{200}$
   Ⓒ 0.05
   Ⓓ $\frac{1}{20}$

6. Mrs. Mosley asked her students to write a group of equivalent numbers. Below are the responses from four of her students.

   - Kasen: 207%, $2\frac{7}{10}$, 2.07
   - Ella: $\frac{12}{5}$, 2.4, 24%
   - Jack: 2.125, 212.5%, $2\frac{1}{8}$
   - Kaitlyn: 2.02, $\frac{101}{50}$, 202%

   Which of the four students completed Mrs. Mosley's task correctly?

   Ⓕ Kasen only
   Ⓖ Ella and Jack only
   Ⓗ Kasen, Ella, and Kaitlyn only
   Ⓙ Jack and Kaitlyn only

7. It rained 30.2% of the days in February. Which number is NOT equivalent to 30.2%?

   Ⓐ $\frac{302}{100}$
   Ⓑ 0.302
   Ⓒ $\frac{302}{1000}$
   Ⓓ $\frac{30.2}{100}$

8. Natalie saved $45 to spend on a new outfit. She bought a sweater for $15. Which shows the part of her savings Natalie spent on the sweater?

   Ⓕ $\frac{1}{2}$
   Ⓖ 30%
   Ⓗ $33\frac{1}{3}$%
   Ⓙ $\frac{3}{5}$

Unit 16 Assessment

Standard 6.4(G) – Readiness

1. Celeste uses 24 inches of ribbon from a spool that holds 36 inches of ribbon. Which shows equivalent representations for the part of the spool that Celeste uses?

   Ⓐ $\frac{1}{3}$, $33\frac{1}{3}\%$, $0.\overline{3}$

   Ⓑ $\frac{2}{3}$, 60%, 0.6

   Ⓒ $0.\overline{6}$, $\frac{2}{3}$, $66\frac{2}{3}\%$

   Ⓓ 30%, 0.3, $\frac{1}{3}$

2. Braden took a survey of his classmates and recorded the information on the chart below.

   Favorite Elective

Elective	Number of Students
Band	15
Art	8
Theater	27
Choir	10

   What part of Braden's classmates chose theater as the favorite elective?

   Ⓕ 0.45   Ⓗ 0.60

   Ⓖ 27%   Ⓙ $\frac{1}{5}$

3. It is estimated that 65.5% of the students at Foster Middle School will attend the benchmark reward party. Which number is NOT equivalent to 65.5%?

   Ⓐ 0.655   Ⓒ 0.0655

   Ⓑ $\frac{655}{1000}$ Ⓓ $\frac{65.5}{100}$

4. While walking in his neighborhood, Quentin found 2 dimes and 2 pennies on the ground. What percent of a dollar did Quentin find?

   Ⓕ 2.2%   Ⓗ 220%

   Ⓖ 22%   Ⓙ 2,200%

5. Shelby spelled 7 out of 10 words correctly on the spelling test. Expressed as a decimal, what part of the words did Shelby spell correctly?

   Record your answer and fill in the bubbles on the grid below. Be sure to use the correct place value.

6. Mrs. Thomas looks at the advertisements from the Sunday paper.

   • Store 1: 30% off
   • Store 2: Save $\frac{2}{5}$
   • Store 3: One-third off
   • Store 4: 40% savings

   Which stores have equivalent savings?

   Ⓕ Store 1 and Store 2

   Ⓖ Store 3 and Store 4

   Ⓗ Store 1 and Store 3

   Ⓙ Store 2 and Store 4

7. Of the sixth graders at Tremont Junior High, $\frac{5}{8}$ plan to take athletics in the seventh grade. Which represents the part of the students who do NOT plan to take athletics in the seventh grade?

   Ⓐ 0.875, 87.5%   Ⓒ 0.375, 37.5%

   Ⓑ 62.5%, 6.25   Ⓓ 58%, 0.58

Standard 6.4(G) — Readiness

Unit 16 Critical Thinking

1. A group of students competed in an art competition. The judges reported each student's score in the table below.

Art Competition

Name	Julie	Seth	Yu	Maya
Score	83%	$\frac{4}{5}$	$\frac{5}{6}$	0.8

Which two students received exactly the same score? _____

Explain how you determined your answer. _____
_____
_____

2. Choose a set of equivalent numbers written as a fraction, decimal, and percent. Use each of the numbers to create three different problems appropriate to that type of number. For example, in problem A, the situation should be best represented by the fractional form of the number.

A. Fraction _____

Problem: _____
_____
_____
_____

B. Decimal _____

Problem: _____
_____
_____
_____

C. Percent _____

Problem: _____
_____
_____
_____

Unit 16 Journal/Vocabulary Activity

Name _____

Standard 6.4(G) – Readiness

Journal

Why is it important to know how to find equivalent forms of numbers (fraction, decimal, and percent)?

_____
_____
_____
_____

Vocabulary Activity

Study the list of numbers. Complete the steps below. Numbers may be used more than once.

$\frac{3}{4}$ 50%   0.6   0.75 $\frac{5}{8}$ 37.5%   3.25 $62\frac{1}{2}\%$

a. Circle one pair of equivalent numbers.

b. Draw a triangle around any number equivalent to 3 parts out of 5.

c. Draw a rectangle around any number equivalent to a fraction in which the numerator is half of the denominator.

d. Place a star below all numbers that mean "per 100."

e. Draw a pentagon around any number greater than 100%.

f. Underline a second pair of equivalent numbers.

g. Draw a heart around all numbers that show how many parts are in the whole.

h. Place a hashtag (#) above all numbers written using base-ten place value.

Name _____

Standard 6.4(G) – Readiness

Unit 16 Motivation Station

Rational Roll

Play *Rational Roll* with a partner. Each pair of players needs two number cubes and a game board. Each player needs a different color pencil and a sheet of paper. Player 1 rolls the cubes and chooses one of the spaces rolled. For example, if player 1 rolls a 1 and 3, he/she may choose (1, 3) or (3, 1). Player 1 records the number from the selected box on his/her paper and generates the other two forms of the number. For example, if a fraction is in the box selected, the decimal and percent forms of the number are generated. If recorded correctly, player 1 then shades that box on the game board with his/her colored pencil, and play passes to player 2. If incorrect, player 2 may "steal" the square by correctly writing all three forms of the number. The winner is the player with more colored squares when time is called or when all squares have been shaded.

	1	2	3	4	5	6
1	20%	$0.\overline{3}$	$\frac{5}{16}$	18.75%	0.9	$\frac{1}{2}$
2	$\frac{5}{8}$	0.875	18%	14.6%	$\frac{1}{4}$	6%
3	0.26	24%	$\frac{3}{10}$	$\frac{2}{5}$	0.8	0.004
4	75%	0.15	0.065	$\frac{2}{3}$	0.7%	$\frac{9}{16}$
5	0.375	$\frac{1}{8}$	60%	$\frac{1}{20}$	0.0625	0.08
6	$\frac{5}{12}$	43.75%	0.1	27%	$\frac{8}{25}$	70%

Unit 16 Homework

Standard 6.4(G) – Readiness

1. Maxine, Sinclaire, and Kyle worked on a science project. They are required to give their calculations in fraction and decimal form. Each person wrote an equation as shown.

   - Maxine wrote $3\frac{1}{5} = 3.2$
   - Sinclaire wrote $5\frac{1}{4} = 5.14$
   - Kyle wrote $4\frac{5}{8} = 4.625$

   Who wrote an equation that is true? Show all work.

2. Anthony has the following coins in his pocket: 3 quarters, 4 dimes, 6 nickels, and 2 pennies. Complete the table by finding what part of a dollar Anthony has of each type of coin. Write all fractions in lowest terms.

   ### Anthony's Coins

Coin	Fraction	Decimal	Percent
Pennies			
Nickels			
Dimes			
Quarters			

Use the table to answer questions 3–5.

Sam surveyed 32 of his classmates to determine their favorite after-school activity. The results of the survey are given in the table.

Favorite After-School Activity

Activity	Part of Votes
Playing video games	$\frac{1}{4}$
Watching television	6 students
Playing sports	0.375
Reading/studying	18.75%

3. What percent of Sam's classmates chose playing video games as their favorite after-school activity?

   _____

   Explain how you found your answer.

   _____
   _____

4. Which activities had an equivalent part of the votes?

   _____

5. What fraction of Sam's classmates chose playing sports as their favorite after-school activity?

   _____

Connections

1. Look around your house for examples of fractions, decimals, and percents. Find at least two examples of how each type of number is used and write them down. What would happen if you switched the numbers to a different form? Would it be easier or harder to use the item? Record your thoughts to share with the class.

2. Play a game of memory with an older sibling or parent. Make a set of 15 cards by choosing five fraction, decimal, and percent equivalents and recording one form on each card. Lay all the cards face down, and take turns turning over three cards to make a match. When all the cards are picked up, the winner is the person with the most matches.

Name _____

Standard 6.4(H) – Readiness

Unit 17 Introduction

1. A bottle of eyedrops holds 15 milliliters of fluid. How is 15 milliliters expressed in liters?

   _____

2. Mrs. Zelenock makes chili for 60 teachers at Boulter Middle School. Her recipe makes enough for 15 servings. If the recipe calls for 3 cups of tomato sauce, how many pints of tomato sauce will Mrs. Zelenock need to purchase in order to serve all the teachers?

   _____

3. An adult female polar bear has an average weight of 435 kilograms. The average weight of an adult male polar bear is 715 kilograms. What is the difference, in grams, of the weights of an adult male and an adult female polar bear?

   _____

4. Mrs. Killian decorates a small square bulletin board in her classroom. The bulletin board has a perimeter of 8 yards. What is the length of one side of the bulletin board, in inches?

   _____

5. Chloe purchased a bag of food pellets for the animals at the petting zoo. Chloe fed the goats 619 grams of food pellets and gave the calves 958 grams. She gave her sister 172 grams of food pellets for the rabbits and 276 grams to feed the chickens. After the feedings, Chloe had 100 grams of food pellets left. How many kilograms did the bag of food pellets weigh before feeding the animals?

   _____

6. Madison purchases 3 gallons of milk. How many one-cup servings are equivalent to 3 gallons?

   _____

7. Juan drives an 18-wheeler transporting 7 new vehicles to a dealership. If the weight per vehicle is 3,500 pounds, write a proportion to find the weight, in tons, of all the vehicles Juan transports.

   _____

   What is the total weight, in tons, of all the vehicles?

   _____

Unit 17 Guided Practice

Standard 6.4(H) – Readiness

1. The recycling center processed 35,000 pounds of recyclable materials. How many tons of recyclable materials are processed?

   Ⓐ $17\frac{1}{2}$ tons

   Ⓑ 70,000,000 tons

   Ⓒ 175 tons

   Ⓓ 16 tons

2. Jenna used the recipe shown below to make punch for a party.

   Fruit Punch Recipe

Ingredient	Amount
Orange juice	4 cups
Pineapple juice	4 cups
Red tropical juice	8 cups
Ginger ale	16 cups

   If Jenna made half a recipe of punch, how many quarts of fruit punch did she make?

   Ⓕ 16 quarts

   Ⓖ 4 quarts

   Ⓗ 32 quarts

   Ⓙ $\frac{1}{2}$ quart

3. Mrs. Stinchfield removed four strips of wallpaper border from her laundry room. The first piece was 110 centimeters long. The second piece was 1.5 meters long. The third piece was 125 centimeters long, and the fourth piece was 205 centimeters long. What was the total length of the wallpaper border Mrs. Stinchfield removed?

   Ⓐ 4.415 meters

   Ⓑ 441.5 meters

   Ⓒ 590 meters

   Ⓓ 5.9 meters

4. Chelsea jogged in the park four days last week as shown in the table below.

   Chelsea's Jogging Record

Day	Distance Jogged
Thursday	3 miles
Friday	2 miles
Saturday	1.5 miles
Sunday	$3\frac{1}{2}$ miles

   What distance, in yards, did Chelsea jog on the weekend?

   Record your answer and fill in the bubbles on the grid below. Be sure to use the correct place value.

5. Which equation is NOT true?

   Ⓐ $\frac{1 \text{ m}}{100 \text{ cm}} = \frac{3 \text{ m}}{300 \text{ cm}}$

   Ⓑ $\frac{1{,}000 \text{ g}}{1 \text{ kg}} = \frac{5{,}200 \text{ g}}{5.2 \text{ kg}}$

   Ⓒ $\frac{1 \text{ L}}{1{,}000 \text{ mL}} = \frac{1.5 \text{ L}}{1{,}500 \text{ mL}}$

   Ⓓ $\frac{10 \text{ cm}}{1 \text{ mm}} = \frac{20 \text{ cm}}{2 \text{ mm}}$

Name _____

Standard 6.4(H) – Readiness

Unit 17 Independent Practice

1 Sasha drinks 2.5 liters of water each weekday. She drinks a total of 3 liters of water on the weekend. What is the total volume of water, in milliliters, Sasha drinks each week?

Ⓐ 15,500 mL
Ⓑ 0.0155 mL
Ⓒ 1,550 mL
Ⓓ 155,000 mL

2 Weston and Zachary collect aluminum cans for three months. The table shows the weights each boy collected.

Aluminum Can Log

Month	Weston	Zachary
1	48 oz	56 oz
2	96 oz	64 oz
3	72 oz	32 oz

The boys combine their cans and take them to the local recycling center at the end of the three months. If the recycling center pays $0.40 per pound for aluminum, how much money will Weston and Zachary earn?

Ⓕ $368
Ⓖ $23
Ⓗ $9.20
Ⓙ $147.20

3 Mrs. Cranford bought a case of bottled water for $5.76 to take to her son's class picnic. The case contained 24 half-pint bottles of water. How many fluid ounces of water did Mrs. Cranford purchase?

Ⓐ 192 fl oz
Ⓑ 138.24 fl oz
Ⓒ 96 fl oz
Ⓓ 384 fl oz

4 Which equation is true?

Ⓕ 500 mL = 5 L
Ⓖ 5 m = 50 cm
Ⓗ 5,000 g = 50 kg
Ⓙ 50,000 mg = 50 g

5 Mrs. Middleton makes a solution to clean her windows. She uses a 2:1 ratio for cups of water to cups of vinegar. If Mrs. Middleton uses a gallon of water to make the window cleaner, how much vinegar does she need?

Ⓐ 12 cups
Ⓑ 2 quarts
Ⓒ 2 pints
Ⓓ 1 gallon

6 Mr. Gregory drives a furniture delivery truck four days each week. The table below shows the driving record for one week.

Weekly Driving Record

Day	Distance Traveled
Monday	30.515 km
Tuesday	105.4 km
Wednesday	74.6 km
Thursday	80.75 km

Find the difference, in meters, between the distances Mr. Gregory traveled on Wednesday and Thursday.

Record your answer and fill in the bubbles on the grid below. Be sure to use the correct place value.

Unit 17 Assessment

Standard 6.4(H) – Readiness

1. Mr. Lopez picks up 2 pallets of dog food to deliver to the animal shelter. Each pallet contains 40 bags of dog food, and the dog food weighs $37\frac{1}{2}$ pounds per bag. What was the total weight, in tons, of the dog food Mr. Lopez delivered to the animal shelter?

    Ⓐ $1\frac{1}{2}$ T

    Ⓑ 2 T

    Ⓒ $2\frac{1}{2}$ T

    Ⓓ $\frac{3}{4}$ T

2. The running path at Rose City Trail is 2.5 kilometers long. The table below shows the number of times Mack ran the trail last week.

    Mack's Running Record

Day	Number
Monday	3
Tuesday	2
Thursday	3

    How many meters did Mack run at Rose City Trail last week?

    Ⓕ 2,000 m

    Ⓖ 20,000 m

    Ⓗ 8 m

    Ⓙ 8,000 m

3. Mr. Enloe uses a 3-foot-long piece of wood to build a square frame. Mr. Enloe cuts the wood in 4 pieces of equal length. What is the length, in inches, of each piece of wood?

    Ⓐ 12 inches

    Ⓑ 9 inches

    Ⓒ 36 inches

    Ⓓ 13.5 inches

4. The doctor prescribes a 10-day antibiotic for Eduardo's ear infection. A dosage of 5 milliliters should be taken twice a day. Expressed as a decimal, how many liters of medicine will Eduardo take in 10 days?

    Record your answer and fill in the bubbles on the grid below. Be sure to use the correct place value.

5. Tabitha used 5 different decorative trims for a craft project. She cut lengths of $5\frac{1}{2}$ inches, 11 inches, $8\frac{1}{2}$ inches, 2 feet, and 35 inches. How much trim did Tabitha use for the craft project?

    Ⓐ $6\frac{1}{2}$ feet

    Ⓒ 7 feet

    Ⓑ 62 inches

    Ⓓ 2 yards

6. Which of the following equations are true?

    I. 12 cm = 0.12 m

    II. 3,400 mm = 34 cm

    III. 5.60 L = 5,600 mL

    IV. 78.9 g = 0.0789 kg

    Ⓕ I only

    Ⓖ III and IV only

    Ⓗ I, III, and IV only

    Ⓙ I, II, and III only

Name _____

Standard 6.4(H) – Readiness

Unit 17 Critical Thinking

1. Allison missed the following problem on her math test.

$$2.4 \text{ kg} = ? \text{ mg}$$

$$1 \text{ kg} = 1{,}000{,}000 \text{ mg}$$

$$\frac{1}{1{,}000{,}000} = \frac{x}{2.4}$$

$$x = 0.0000024 \text{ mg}$$

Describe the mistake or mistakes Allison made on the problem. Then work the problem correctly.

_____
_____
_____
_____

2. Select one category of the customary measurement system (length, weight, or capacity), and write a rap, poem, or other memory tool to help you remember the relationships in that category. Present your creation to the class.

_____
_____
_____
_____
_____
_____
_____
_____
_____
_____

Unit 17 Journal/Vocabulary Activity

Name _____

Standard 6.4(H) – Readiness

Journal

You have been learning to make conversions within the customary and metric measurement systems. In which system do you think it is easier to make conversions? Justify your answer.

_____
_____
_____

Vocabulary Activity

Complete the concept web below for the following metric system prefixes: *milli-*, *centi-*, *deci-*, *deca-*, *hecto-*, and *kilo-*. List at least three words that begin with the prefix. Include the metric measurement word in your examples.

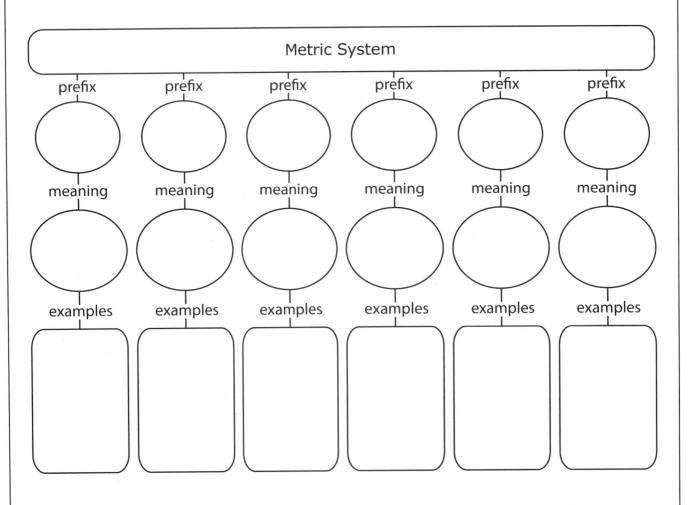

Name _____

Standard 6.4(H) — Readiness

Unit 17 Motivation Station

Tic-Tac-Conversion

Play *Tic-Tac-Conversion* with a partner. Each pair of players needs a paper clip to use with the spinner, a game board, and a deck of playing cards with all face cards (Jacks, Queens, and Kings) removed. Each player needs a sheet of paper, a pencil, a measurement conversion chart, and colored tokens. For this game, Ace equals one. Player 1 spins the spinner and flips over the top card. He/she uses the number on the card to make the conversion described on the spinner. For example, if "yards to feet" is spun and the number 10 shows on the card, then 10 yards are converted to feet. If correct, player 1 places a token on the grid. If incorrect, play passes to player 2 who repeats the process. The game ends when one player connects 4 spaces in a row on the grid, either vertically, horizontally, or diagonally.

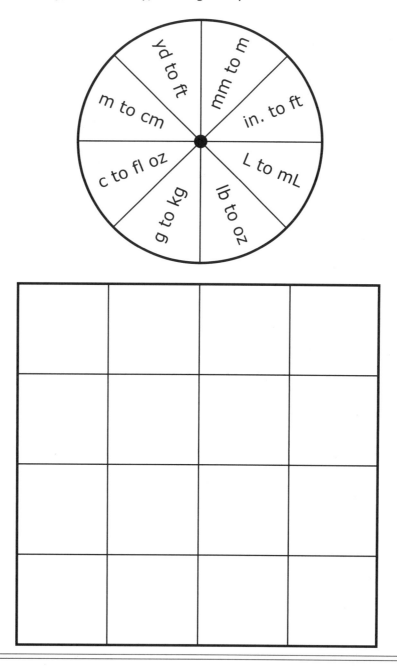

©2014 mentoringminds.com   motivation**math**™ LEVEL 6   ILLEGAL TO COPY   141

Unit 17 Homework

Standard 6.4(H) – Readiness

1 Kelsey makes scented candles to sell at the local craft fair. Each candle uses 10 ounces of wax and 1 ounce of scent. Explain in complete sentences how Kelsey can determine the number of candles that can be made if she has 50 pounds of wax.

_____

_____

How many candles will Kelsey make from 50 pounds of wax? Show all work.

_____

2 For his science fair project, Michael watered his bean plants different amounts of water each week. The table shows the amount of water each plant received.

Michael's Science Project

Plant Number	Amount of Water
1	325 mL
2	570 mL
3	287 mL
4	608 mL

Expressed as a decimal, how many total liters of water did Michael's plants receive during the six weeks before the science fair? Show all work.

_____

3 Jill wants to make lemonade for her family reunion. She needs enough lemonade for 60 people to each have 1 cup. Write a proportion Jill could use to find the number of gallons of lemonade she will need to make.

How many gallons of lemonade does Jill need to make? Show all work.

_____

4 Mrs. Tiff wrote the following problem on the board.

Brandon drank two 2-liter bottles of soda. How many milliliters did Brandon drink?

Louis got an answer of 4,000 milliliters. Miranda got 0.004 milliliters. Explain how each student got his/her answer.

_____

_____

Which student is correct? Why?

_____

_____

5 It is approximately 32 miles from Dallas to Fort Worth. Write and solve a number sentence that could be used to find f, the number of feet from Dallas to Fort Worth.

Connections

1. Have a scavenger hunt with your family. Make a list of different weights and measurements (ounces, cups, feet, milliliters, etc.), and search the house for items that match. Once you have found everything on the list, explain to your family how to make conversions. For example, if a gallon of milk is on the list, explain how 1 gallon converts to 4 quarts or 8 pints.

2. Create a poster of objects used to measure. Divide the poster into three sections and label the sections Weight, Length, and Capacity. Find pictures of measurement tools used for each type of measurement, and tape them in the correct section. Share your poster with the class.

Standard 6.5(A) – Supporting

Unit 18 Introduction

Use the graph to answer questions 1–4.

The graph below shows the cost for different numbers of pencils from the school store.

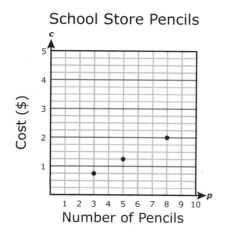

School Store Pencils

1. Write and solve a proportion to determine c, the cost of 12 pencils.

2. How many pencils can you purchase from the school store with $5?

3. Determine the unit cost of a pencil using the graph. Use the unit cost to write a multiplicative equation to find c, the cost of 15 pencils.

4. Complete the table shown below.

School Store Pencils

Number of Pencils	Total Cost
	$0.50
7	
	$2.75
16	

5. Jacob runs 2 miles in 16 minutes 15 seconds. Write a proportion to determine t, the time in minutes it takes Jacob to run $\frac{1}{2}$ mile at the same pace.

Use the graph to answer questions 6 and 7.

The graph shows the multiplicative relationship between x and y.

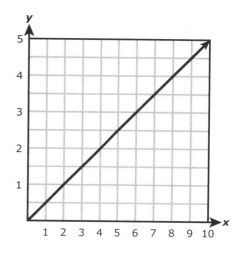

6. What factor would be used to determine the value of y if $x = 12$?

7. Determine the value of x when $y = 12.5$.

Unit 18 Guided Practice

Standard 6.5(A) – Supporting

1 A recipe for a dozen iced cookies calls for $2\frac{1}{4}$ cups flour. Which can be used to determine the amount of flour needed to make 32 cookies?

Ⓐ $\dfrac{2\frac{1}{4}}{12} = \dfrac{32}{x}$ Ⓒ $\dfrac{2\frac{1}{4}}{1} = \dfrac{32}{x}$

Ⓑ $\dfrac{2\frac{1}{4}}{1} = \dfrac{x}{32}$ Ⓓ $\dfrac{2\frac{1}{4}}{12} = \dfrac{x}{32}$

Use the graph to answer questions 2 and 3.

The graph shows the multiplicative relationship between inches and feet.

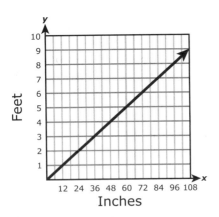

2 What is the relationship between x and y?

Ⓕ $y = 12x$ Ⓗ $y = \frac{1}{6}x$

Ⓖ $y = 6x$ Ⓙ $y = \frac{1}{12}x$

3 Expressed as a decimal, find the number of feet in 66 inches.

Record your answer and fill in the bubbles on the grid below. Be sure to use the correct place value.

4 The ratio of green tiles to purple tiles in the art class mosaic is 5 to 8. There are 45 green tiles in the mosaic. Which shows a way to find the number of purple tiles in the mosaic?

Ⓕ $8 + 40 = 48$

Ⓖ $45 \div 5 = 9$

Ⓗ $8(9) = 72$

Ⓙ $5(8) = 40$

Use the table to answer questions 5 and 6.

A catering company mixes water with fruit-flavored powdered mix to create punch. The recipe calls for 1 cup of water for every 0.5 cup of powdered mix. The table shows the amounts of water, in cups, for various amounts of powdered mix.

Water (c)	$\frac{1}{2}$	1	$2\frac{1}{2}$	6
Mix (c)		$\frac{1}{2}$		3

5 How much powdered mix will the catering company need to mix with $2\frac{1}{2}$ cups of water?

Ⓐ $1\frac{1}{2}$ cups

Ⓑ 1.45 cups

Ⓒ $1\frac{2}{5}$ cups

Ⓓ $1\frac{1}{4}$ cups

6 Based on the values in the table, which expression can be used to calculate how much water is needed to mix with $4\frac{3}{4}$ cups of fruit-flavored powdered mix?

Ⓕ $4.75 \times \dfrac{1}{0.5}$

Ⓖ $1 \times \dfrac{0.5}{4.75}$

Ⓗ $4.75 \times \dfrac{0.5}{1}$

Ⓙ $0.5 \times \dfrac{1}{4.75}$

Name _____

Standard 6.5(A) – Supporting

Unit 18 Independent Practice

1. Nathaniel can run 3 laps in the gym in 2 minutes. If Nathaniel continues to run at the same rate, which of the following graphs best represents his progress?

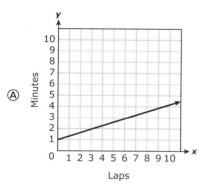

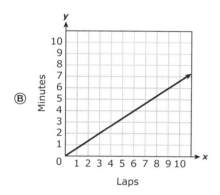

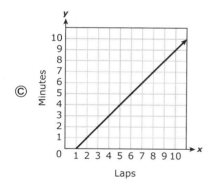

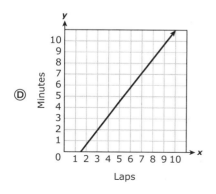

Use the table to answer questions 2 and 3.

The table shows the relationship between the amount of time a bricklayer works and the height of the wall he builds.

Wall Height

Time Worked (hours)	Wall Height (inches)
2	12
3	18
5	30

2. If the bricklayer works for 9 hours, what would be the height of the wall in inches?

Ⓕ 42 in. Ⓗ 60 in.
Ⓖ 56 in. Ⓙ Not here

3. Which expression can be used to determine the number of hours it takes the bricklayer to reach a height of w inches?

Ⓐ $6w$ Ⓒ $12w$
Ⓑ $\frac{1}{2}w$ Ⓓ $\frac{1}{6}w$

4. In keyboarding class, Josephine can type 55 words per minute. If she were to continue at this speed, which proportion could be used to determine x, the number of words typed in 1.5 hours?

Ⓕ $\frac{x}{1.5} = \frac{55}{1}$ Ⓗ $\frac{55}{90} = \frac{x}{1}$
Ⓖ $\frac{1}{55} = \frac{90}{x}$ Ⓙ $\frac{1}{55} = \frac{x}{1.5}$

5. Nicole purchased 6 pounds of cheese for $6.90 from the deli. If the amount of cheese, in pounds, is changed by a factor of $\frac{1}{3}$, which expression can be used to determine the cost?

Ⓐ $\frac{1}{3}(6)$ Ⓒ $3(6)$
Ⓑ $3(6.90)$ Ⓓ $\frac{1}{3}(6.90)$

Unit 18 Assessment

Standard 6.5(A) – Supporting

1 Hallie reads 6 pages from her book in 8 minutes. If she continues reading at the same rate and the amount of time changes by a factor of $\frac{5}{2}$, how many pages will Hallie read?

Ⓐ 15 pages Ⓒ 20 pages
Ⓑ 120 pages Ⓓ 3 pages

Use the graph to answer questions 2 and 3.

The graph represents the cost of bakery cookies.

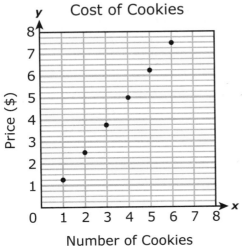

2 What is the cost of 11 cookies?

Record your answer and fill in the bubbles on the grid below. Be sure to use the correct place value.

3 Which equation could NOT be used to determine the cost of two dozen cookies?

Ⓐ $\frac{3}{3.75} = \frac{24}{x}$ Ⓒ $\frac{5}{4} = \frac{x}{24}$

Ⓑ $\frac{x}{24} = \frac{5}{6.25}$ Ⓓ $\frac{24}{x} = \frac{2}{2.5}$

4 Mr. Brock pays $780 per year for his gym membership. Which table best represents the relationship between m, the number of months and f, the amount Mr. Brock pays for that length of time?

Ⓕ
m	$f(\$)$
1	65
4	260
6	360

Ⓗ
m	$f(\$)$
3	165
9	585
12	780

Ⓖ
m	$f(\$)$
2	130
8	520
10	650

Ⓙ
m	$f(\$)$
5	325
7	385
11	715

5 Kenya is running for student council president. She can paint 2 posters in 15 minutes. If she continues at this rate, which of the following methods could be used to determine the number of posters Kenya will paint in $1\frac{3}{4}$ hours?

Ⓐ Multiply the number of posters by 7.5

Ⓑ Solve the proportion $\frac{2}{15} = \frac{x}{1.75}$

Ⓒ Multiply the number of posters by 7

Ⓓ Solve the equation $105\left(\frac{15}{2}\right) = x$

6 There are 16 pink jelly beans in a bag. The ratio of red to pink jelly beans in the bag is $\frac{3}{4}$. Which proportion could NOT be used to find r, the number of red jelly beans in the bag?

Ⓕ $\frac{3}{4} = \frac{r}{16}$ Ⓗ $\frac{r}{3} = \frac{16}{4}$

Ⓖ $\frac{16}{r} = \frac{4}{3}$ Ⓙ $\frac{3}{16} = \frac{4}{r}$

Name _____

Standard 6.5(A) – Supporting

Unit 18 Critical Thinking

1. John makes trail mix to take on his hike. For every cup of raisins, he adds 2 cups of mixed nuts. John creates the graph below to represent this ratio. Complete the table using data from the graph.

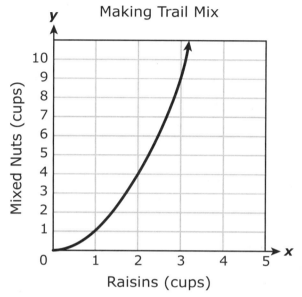

Making Trail Mix

Raisins (cups)	Mixed Nuts (cups)
1	
2	
3	

Do the table and graph correctly represent the ratio of raisins to mixed nuts in John's trail mix? Explain.

_____

If the graph does NOT represent the ratio of raisins to mixed nuts in John's trail mix, correctly graph the relationship on the same graph above. What points did you use to create your graph?

_____

If John adds 12 cups of mixed nuts to his trail mix, how many cups of raisins should he add?

_____

2. Write a word problem to show how to use a proportion to represent a ratio or rate and solve a problem.

_____
_____
_____
_____

Name _____

Unit 18 Journal/Vocabulary Activity

Standard 6.5(A) – Supporting

Journal

Explain how to find the actual distance between two points on a map when you know the scale used to create the map is 1 inch equals 5 miles.

_____
_____
_____

Vocabulary Activity

The box below contains multiple examples of ratios, rates, unit rates, and proportions. Sort the examples by writing each in the most descriptive bubble. Under each bubble, justify your placement of the examples.

2 dogs : 3 cats	3 cans for $1	$\dfrac{360 \text{ miles}}{20 \text{ gallons}}$
$\dfrac{3 \text{ wins}}{45 \text{ total}} = \dfrac{w}{15 \text{ total}}$	65 miles per hour	150 students to 5 teachers
	25 widgets every 15 minutes	$\dfrac{2 \text{ bags}}{3 \text{ days}} = \dfrac{b}{30 \text{ days}}$

Ratios

Justify: _____

Rates

Justify: _____

Unit Rates

Justify: _____

Proportions

Justify: _____

Standard 6.5(A) – Supporting

Unit 18 Motivation Station

The picture below has a grid overlay made up of $\frac{1}{2}$-inch squares. Transfer the picture to the larger blank grid, made up of $1\frac{1}{2}$-inch squares. What is the factor used to transfer the smaller picture to the larger picture?

Unit 18 Homework

Standard 6.5(A) – Supporting

1 To make 120 cupcakes, Glenn used 5 boxes of cake mix. Based on this rate, complete the graph below.

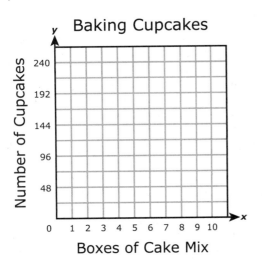

How many cupcakes will Glenn make using 7 boxes of cake mix?

_____

2 At the craft fair, Gracie sold 18 bracelets in 4.5 hours. Write and solve a proportion to find the number of bracelets, b, Gracie sold each hour.

3 There are 760 students at Wyatt Middle School. On Thursday, 500 students attended the pep rally after school. Write a proportion that could be used to represent the number of sixth-grade students attending the pep rally if there are 266 sixth-grade students at Wyatt Middle School.

_____

How many sixth-grade students attended the pep rally? Show all work.

_____

4 The ratio of students that prefer pepperoni pizza to sausage pizza at Adams Junior High is 5 to 3. Complete the table below to represent the numbers of students that would be expected to choose pepperoni and sausage pizza if there are 640 students at Adams Junior High.

Pizza Choices

Pepperoni	Sausage	Total Students
5	3	8
10		
	30	
		240
	180	
350		
		640

How many students at Adams Junior High would be expected to choose pepperoni pizza?

_____

Connections

1. Locate a linear graph in a textbook from another subject area. Briefly explain what the graph is about and identify the ratio or rate the graph represents. Share the information with the class.

2. Ask a parent or grandparent for a copy of a recipe for your favorite dish. Scale the recipe up to serve the entire class. What will the scale factor be? Calculate the amount of each ingredient that will be needed to make the recipe using the scale factor.

Name _____

Standard 6.5(B) – Readiness

Unit 19 Introduction

1. Marcela is shopping for a new winter coat. She finds a coat that is regularly priced at $60 on sale for 25% off. Use the model below to determine how much money Marcela saves by purchasing the coat on sale.

$5	$5	$5
$5	$5	$5
$5	$5	$5
$5	$5	$5

How much money does Marcela save?

_____

How much money does Marcela pay for the coat?

_____

2. Lucy correctly answered 70% of the questions on her science homework. She counted 14 correct answers on her paper. If each question is of equal value, how many questions did Lucy have on her science homework?

_____

Draw a strip diagram to represent your solution.

3. The school cafeteria offered a choice of pizza or nachos to sixth-grade students. Seventy-seven students chose pizza. This represented 35% of the sixth-grade class. How many students are in the sixth grade?

_____

4. Coach West placed 28 of his athletes on suspension due to failing a class. If he has 200 athletes, explain how the model below could be used to determine the percent of athletes Coach West placed on suspension.

_____
_____

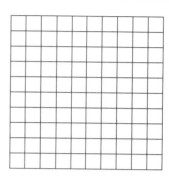

What percent of the athletes did Coach West suspend?

_____

5. Mrs. Martinez assigned 20 pages of reading for homework. Jace had time to read 3 pages before he left school. Use the number lines below to determine the percent of the assigned pages Jace read while at school.

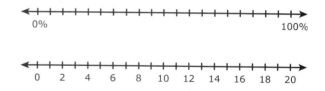

Write a proportion that can be used to determine the percent of pages Jace read at school.

_____

What percent of the pages does Jace need to read at home?

_____

Unit 19 Guided Practice

Standard 6.5(B) – Readiness

1. This year, Mrs. Fisher has 9 students who are new to the school. The number of new students represents 7.5% of all her students. Which equation could be used to determine x, the total number of students Mrs. Fisher has this year?

   Ⓐ $\frac{7.5}{100} = \frac{x}{9}$

   Ⓑ $\frac{7.5}{9} = \frac{x}{100}$

   Ⓒ $\frac{x}{7.5} = \frac{9}{100}$

   Ⓓ $\frac{9}{x} = \frac{7.5}{100}$

2. The shaded model below represents the percent of students at Judson Middle School who bring lunches to school.

   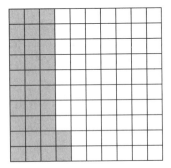

   If 650 students attend Judson Middle School, how many bring lunches?

   Ⓕ 32

   Ⓖ 208

   Ⓗ 442

   Ⓙ 618

3. Natalie saved $20 when she purchased a new phone. The phone originally cost $125. What percent savings did Natalie receive on the purchase of the new phone?

   Ⓐ 20%

   Ⓑ 16%

   Ⓒ 6.25%

   Ⓓ 25%

4. Kareem correctly answered 85% of the questions on his last math test. If he missed 6 problems, how many questions were on the test?

   Ⓕ 40

   Ⓖ 70

   Ⓗ 20

   Ⓙ 30

5. Berkley surveyed students in his fourth-period class to determine how they travel to school each morning.

   School Transportation

Transportation	Number of Students
Walk	3
Bus	14
Car	8

   What percent of the students Berkley surveyed ride the bus to school?

   Ⓐ 14%

   Ⓑ 32%

   Ⓒ 56%

   Ⓓ Not here

6. Leticia wants to go on a band trip to New York for spring break. It costs $1,200 per student. The band will pay 25% of the cost for each student.

   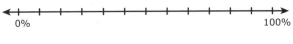

   Using the number lines, determine the amount of money Leticia will pay for the New York trip.

   Ⓕ $300

   Ⓖ $900

   Ⓗ $400

   Ⓙ $800

Standard 6.5(B) – Readiness

Unit 19 Independent Practice

1. Of the 350 songs D'Marco downloaded, 24% were free. What is the total number of free songs D'Marco downloaded?

   Ⓐ 14 Ⓒ 326
   Ⓑ 84 Ⓓ Not here

2. Liam practices free throws every morning before he goes to school. This morning, Liam made 42 out of 50 free throws. Which equation can be used to find x, the percent of free throws Liam made?

   Ⓕ $\frac{42}{100} = \frac{x}{50}$
   Ⓖ $\frac{8}{50} = \frac{x}{100}$
   Ⓗ $\frac{100}{x} = \frac{42}{50}$
   Ⓙ $\frac{42}{50} = \frac{x}{100}$

3. During a school assembly, 15 out of 60 students wore orange shirts to show school spirit. What percent of the students did NOT wear orange shirts?

   Ⓐ 8%   Ⓒ 25%
   Ⓑ 15% Ⓓ 75%

4. Tatiana ordered a floral arrangement with 60% roses. The florist told her she would put 15 roses in the arrangement. How many total flowers should Tatiana expect in the floral arrangement she ordered?

   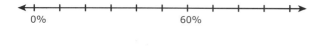

   Ⓕ 20   Ⓗ 28
   Ⓖ 25   Ⓙ 23

5. Landon and Ella ate dinner at Billy's Barbeque. The cost of dinner was $25.50. They left the waiter a 20% tip. How much money did Landon and Ella leave for the tip?

   Record your answer and fill in the bubbles on the grid below. Be sure to use the correct place value.

6. Brady recorded the number and color of cars in the parking lot. The shaded model represents the percent of white vehicles counted in the parking lot.

   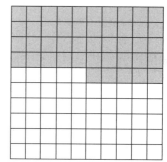

   If Brady counted a total of 81 white cars, how many cars did he count in the parking lot?

   Ⓕ 180
   Ⓖ 100
   Ⓗ 126
   Ⓙ 36

Unit 19 Assessment

Standard 6.5(B) – Readiness

1 Mrs. Weatherford spends 15% of each weekday traveling in her car. How many hours per day does Mrs. Weatherford spend in her car during the week?

Ⓐ 3 hours

Ⓑ 3.6 hours

Ⓒ 4 hours

Ⓓ 1.8 hours

2 Six of the 16 students in Mr. Rigsby's classroom after school are serving detention for tardies. The rest of the students are there for tutoring. Using the grid below, what percent of the students in Mr. Rigsby's classroom after school are there for tutoring?

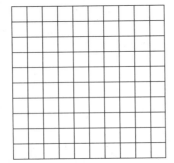

Ⓕ 16%

Ⓖ 37.5%

Ⓗ 60%

Ⓙ 62.5%

3 Mrs. Loya deposits 20% of each weekly paycheck into her savings account. Last week, Mrs. Loya deposited $170 into her savings account. How much money did Mrs. Loya receive in her paycheck last week?

Ⓐ $850

Ⓑ $340

Ⓒ $1,170

Ⓓ $805

4 Logan spends 26% of his weekly allowance on 2 banana splits. Banana splits cost $3.25 each at Frozen Treats. How much money does Logan receive for his weekly allowance?

Record your answer and fill in the bubbles on the grid below. Be sure to use the correct place value.

5 Raj bought a pair of basketball shoes for $70. The sales tax on the shoes was $5.60. Which represents the sales tax rate?

Ⓐ 0.08%   Ⓒ 9%

Ⓑ 12.5%   Ⓓ 8%

6 The current dimensions of Mr. Meyer's garden are shown in the model below.

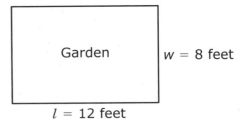

Mr. Meyer plans to increase the width of his garden by 25%. Which equation can be used to determine how many feet Mr. Meyer will add to the width of his garden?

Ⓕ $\frac{25}{100} = \frac{x}{12}$ Ⓗ $\frac{25}{100} = \frac{8}{x}$

Ⓖ $\frac{12}{x} = \frac{25}{100}$ Ⓙ $\frac{x}{8} = \frac{25}{100}$

Name _____

Standard 6.5(B) — Readiness

Unit 19 Critical Thinking

1 In Mrs. Dudley's sixth-grade class, each student was challenged to read 75 books by the end of the school year. Lamar read 60 books by winter break. He read an additional 30 books by the end of the year. Sketch a model below to find the percent of the challenge Lamar met by the end of the school year.

What percent of Mrs. Dudley's challenge did Lamar meet? _____

2 Sandra bought a pair of boots that cost $75. After tax, her total cost was $82.50. Explain how to find the tax rate Sandra paid on her boots.

_____

_____

_____

What was the tax rate Sandra paid on her boots? _____

Unit 19 Journal/Vocabulary Activity

Standard 6.5(B) – Readiness

Journal

The picture shows the progress of a local high school's fund-raising efforts.

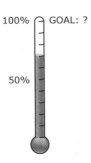

The school has raised $157,500 so far. Explain how to use the picture to find the school's goal.

_____
_____

Vocabulary Activity

For each of the situations given, describe the part and the whole. An example is done for you.

A bakery used to sell doughnuts for $0.75 each. They increased the price by $0.06 each. By what percent did the bakery increase the cost of a doughnut?

  Part: the increase in price

  Whole: the original cost of the doughnut

1. A belt cost $6.95. The tax rate is 8%. How much tax is added?

   Part: _____

   Whole: _____

2. Sara scored 80% on her last history test. She answered 16 questions correctly. How many questions were on the test?

   Part: _____

   Whole: _____

3. There are 36 people on the set construction crew for the school play. Last Friday, 7 crew members were absent with the flu. What percent of the crew was absent last Friday?

   Part: _____

   Whole: _____

4. In 2012, the Texas Rangers won 57.4% of their regular season games. The Rangers won 93 games. How many games did the Rangers play during the regular season in 2012?

   Part: _____

   Whole: _____

Standard 6.5(B) – Readiness

Unit 19 Motivation Station

On a Roll

Play *On a Roll* with a partner. Each pair of players needs a number cube and a game board. Each player needs a pencil. In turn, players roll the number cube and solve the corresponding problem. Players show work in the space provided and explain how they arrived at their answer. If correct, the player is awarded the number of points in the banner next to the problem. If incorrect, the other player can claim the points by working the problem correctly. If a player rolls the number of a problem that has already been solved, play passes to the next player. The winner is the player with more points after all problems have been solved.

Crystal won 84% of the skating competitions she entered last year. If Crystal entered 250 competitions, how many did she win? Answer _____ **10**	On one day in February, 3 out of 18 spider monkeys at the zoo were ill. What percent of the monkeys were ill? Round your answer to the nearest tenth. Answer _____ **10**
Luke works in the produce department at the local grocery store. On Friday, 35 people bought a pumpkin. If this number is 28% of the total number of customers, how many customers were in the store on Friday? Answer _____ **20**	A total of 112 students voted for class favorite. Denny received $62\frac{1}{2}$% of the vote. How many students voted for Denny for class favorite? Answer _____ **20**
During the school year, Tiffany received 12 referrals for being tardy to class. If these referrals represented 30% of all the referrals Tiffany received during the year, how many referrals did Tiffany receive during the school year? Answer _____ **30**	The girls' basketball team lost 7 out of 25 games this season. What percentage of games did the team win this season? Answer _____ **30**

Unit 19 Homework

Standard 6.5(B) – Readiness

1 Shelby purchased a coat on sale for 30% off the original price of $145. Using the strip diagram, find the amount of the discount.

0%   100%

Discount: _____

2 There are 225 students at March Middle School. On Friday, 135 students wore spirit shirts. What percent of the students did NOT wear spirit shirts on Friday?

_____

3 On Thursday, 72% of the customers at a frozen yogurt shop ordered vanilla frozen yogurt. If 180 customers ordered vanilla frozen yogurt, how many customers visited the frozen yogurt shop on Thursday? Use the model below to help find the answer.

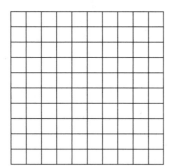

Number of customers: _____

4 At a yard sale, Celia bought an electric guitar for 85% of its original value. Celia paid $75.65 for the guitar. What was the original value of the guitar?

_____

5 The Wiggins family ate dinner at a restaurant. The total cost of the meal was $85.60. The family left a 15% tip for the waiter. Use the number lines to determine the amount of the tip the Wiggins family left the waiter.

Tip amount: _____

6 Amy bought a shirt for $25. Sue bought the same shirt on sale one week later for $10 less. What percent of the original cost did Sue save by purchasing the shirt on sale?

_____

Connections

1. Practice finding discounts while shopping. Find different items that are marked with a percent discount. Use a calculator to find out what the discount will be. Then calculate how much the item will cost after the discount is taken.

2. Look through newspapers and magazines to find examples of percent. Find examples where the percent and the whole or the percent and the part are given. Then, find the missing value. Bring one or two examples to share with the class.

Standard 6.5(C) – Supporting

Unit 20 Introduction

Use the table to answer questions 1–5.

Mr. Smith grows a variety of vegetables in his garden as indicated in the table below.

Vegetable Garden: 120 ft^2

Vegetable	Part of Garden
Okra	$\frac{7}{20}$
Potato	15%
Squash	$\frac{1}{5}$
Tomato	0.2
Bell pepper	10%

1. Expressed as a decimal, what part of Mr. Smith's garden contains bell peppers?

   _____

2. What percent of Mr. Smith's garden does NOT contain okra?

   _____

3. Which two vegetables cover the same area in Mr. Smith's garden?

   _____

4. Which vegetable covers an area of 18 square feet in Mr. Smith's garden?

   _____

5. Mrs. Ray lives next door and has a garden with an area of 180 ft^2. She plants okra in 35% of her garden. Whose garden has more square feet planted with okra? Explain your answer.

   _____

   _____

6. Mr. Smith decides to add a second garden to his property as shown in the table below.

   Vegetable Garden #2: 100 ft^2

Vegetable	Area (ft^2)
Jalapeño	
Cucumber	
Tomato	

   Use the clues below to complete the table.

   Clue 1: The area planted with tomatoes is $\frac{9}{20}$ of the area of the second vegetable garden.

   Clue 2: The section planted with cucumbers is 25% of the area of the second garden.

7. Arturo completed 9 questions from his math homework during class. His twin brother, Raul, completed 40% of the same math homework assignment during class. If there were 30 questions on the math homework assignment, which brother completed more of the assignment in class? Explain your answer.

   _____

   _____

   _____

8. Kylie saved $550 from babysitting. She used 32% of her money for a down payment on summer camp and $\frac{3}{20}$ of her savings for two new video games. She deposited the rest into a college savings account. Expressed as a decimal, what part of her money did Kylie deposit into her savings account?

   _____

Unit 20 Guided Practice

Standard 6.5(C) – Supporting

Use the table to answer questions 1–4.

The table shows Vivian's monthly budget for her income of $2,200.

Monthly Budget

Expense	Part of Income
Rent	30%
Utilities	$\frac{1}{4}$
Transportation	0.1
Groceries	20%
Entertainment	10%
Other	$\frac{1}{20}$

1 Vivian budgets $550 for which expense?

  Ⓐ Rent 　　Ⓒ Groceries
  Ⓑ Utilities 　Ⓓ Other

2 What fraction represents the part of Vivian's income she does NOT budget for rent and utilities?

  Ⓕ $\frac{9}{20}$ 　　Ⓗ $\frac{11}{20}$
  Ⓖ $\frac{1}{2}$ 　　Ⓙ $\frac{12}{25}$

3 Expressed as a decimal, what part of Vivian's budget is allotted for other expenses?

Record your answer and fill in the bubbles on the grid below. Be sure to use the correct place value.

4 Vivian's friend, Skylar, spends $\frac{1}{5}$ of her monthly budget on groceries. Which statement is true about the amount of money both Vivian and Skylar budget on groceries?

  Ⓕ Skylar budgets more money for groceries than Vivian.
  Ⓖ Vivian budgets more money for groceries than Skylar.
  Ⓗ Both Vivian and Skylar budget the same amount for groceries.
  Ⓙ Not enough information is given.

5 Lilly baked two dozen cookies. She took $\frac{3}{8}$ of the cookies to neighbors, shared 12 cookies with her family, and kept the rest. What percent of the cookies did Lilly keep for herself?

  Ⓐ 3%
  Ⓑ 12.5%
  Ⓒ 25%
  Ⓓ 20.5%

6 Montrel uses the following coins to make exact change for his purchase: 2 quarters, 3 dimes, 1 nickel, and 4 pennies. Which statement correctly represents the coins Montrel uses to make exact change?

  Ⓕ Fifty percent of the coins he uses are quarters.
  Ⓖ Two-thirds of the coins he uses are pennies.
  Ⓗ Three-tenths of the coins he uses are dimes.
  Ⓙ One percent of the coins he uses are nickels.

Standard 6.5(C) – Supporting

Unit 20 Independent Practice

1. Of the 25 students in Mrs. Miller's class, $\frac{3}{5}$ made a passing grade on the homework assignment, 0.32 made a failing grade on the homework assignment, and 8% used a homework pass for the assignment. Which statement is NOT true about the completion of homework for students in Mrs. Miller's class?

   Ⓐ Sixty-eight percent of the students did not fail the homework assignment.

   Ⓑ Only 3 students used a homework pass.

   Ⓒ The fraction of students who failed the homework assignment is $\frac{8}{25}$.

   Ⓓ Sixty percent of the students made a passing grade on their homework.

2. Graciela is working to earn $200. She earns $25 mowing her grandmother's lawn and $60 for a day of housecleaning. Which decimal represents the part of the money Graciela still needs to earn?

   Ⓕ 0.575

   Ⓖ 0.115

   Ⓗ 0.425

   Ⓙ 0.85

3. Billy's soccer team won 17 out of the 20 games played. Sam's soccer team lost 3 games of their 12-game season. Which statement is NOT true about the wins and losses for both Billy's and Sam's soccer teams?

   Ⓐ Billy's soccer team won 17 games which represents 85% of their games.

   Ⓑ Sam's soccer team won 9 games which represents $\frac{3}{4}$ of their games.

   Ⓒ Both of their teams lost 3 games, so their losing percentage would be the same.

   Ⓓ Sam's soccer team lost 3 games which represents 25% of their games.

Use the table to answer questions 4 and 5.

Using a bag of fruit snacks, Summer records the information shown in the table below.

Fruit Snacks

Color	Count	Calculations
Red	4	$\frac{1}{4} = 25\%$
Yellow	3	$\frac{3}{16} = \frac{18.75}{100}$
Orange	2	$\frac{1}{8} = 0.0125$
Green	7	$\frac{7}{16} = 43.75\%$

4. Which color fruit snack shows an incorrect calculation?

   Ⓕ Red, because $\frac{4}{16} = 0.25$

   Ⓖ Yellow, because $\frac{3}{16} = 18.75\%$

   Ⓗ Orange, because $\frac{2}{16} = 12.5\%$

   Ⓙ Green, because $\frac{7}{16} = 0.4375$

5. If Summer eats a red fruit snack and recalculates the information, will the percentage of yellow fruit snacks change?

   Ⓐ No, because the number of yellow fruit snacks does not change

   Ⓑ Yes, because yellow now represents $\frac{2}{15}$ of the fruit snacks

   Ⓒ No, because the total number of fruit snacks does not change

   Ⓓ Yes, because yellow now represents 20% of the fruit snacks

Unit 20 Assessment

Name _____

Standard 6.5(C) — Supporting

1. Cooper bought a box of $2\frac{1}{2}$ dozen donuts. The box contained $\frac{3}{10}$ glazed, 6 chocolate, 0.4 pink sprinkles, and 10% cinnamon twists. Which statement does NOT correctly represent the donuts Cooper purchased?

   Ⓐ Seventy percent of the donuts Cooper bought were not glazed.

   Ⓑ Cooper bought 2 cinnamon twists.

   Ⓒ Two-tenths of the donuts Cooper bought were chocolate.

   Ⓓ Cooper bought a dozen pink sprinkle donuts.

2. Mrs. Ulrich bought 2 large take-and-bake pizzas. She cut one pizza into 12 equal slices for her children. She cut the other pizza into 8 equal slices for herself and her husband. Mrs. Ulrich and her husband ate 75% of their pizza. The children ate 9 slices of their pizza. Who ate the greater part of their pizza?

   Ⓕ Mr. and Mrs. Ulrich, because 75% = $\frac{6}{8}$

   Ⓖ The children, because 9 slices is greater than 6 slices

   Ⓗ The parents and the children ate equal parts of their pizzas, because $\frac{9}{12}$ = 75%

   Ⓙ Not here

3. In the last inning of the game, Ethan threw 12 strikes out of 18 pitches. Which statement best describes the pitches Ethan threw in the last inning?

   Ⓐ Of the pitches Ethan threw, $66\frac{2}{3}$% were strikes.

   Ⓑ Ethan's strike ratio is 4 to 9.

   Ⓒ Ethan did not pitch strikes 30% of the time.

   Ⓓ The part of Ethan's pitches that were strikes can be represented by 0.6.

Use the table to answer questions 4–6.

Talia's Girl Scout troop sells cookies every year. Their troop leader set a goal for each girl to sell 150 boxes of cookies. The table shows the part of the goal that was met by each troop member.

Cookie Sales

Troop Member	Part of Goal Met
Talia	0.6
Vickie	$\frac{19}{25}$
Trinity	54%
Zana	$\frac{4}{5}$
Shameka	104%

4. What fraction of the goal did Shameka meet?

   Ⓕ $\frac{21}{25}$ Ⓗ $\frac{75}{71}$

   Ⓖ $\frac{31}{30}$ Ⓙ $\frac{26}{25}$

5. Expressed as a decimal, what part of the goal did Vickie meet?

   Record your answer and fill in the bubbles on the grid below. Be sure to use the correct place value.

6. Who sold 81 boxes of cookies?

   Ⓕ Talia   Ⓗ Vickie

   Ⓖ Trinity   Ⓙ Zana

Name _____

Standard 6.5(C) – Supporting

Unit 20 Critical Thinking

1. The following models represent three equivalent values.

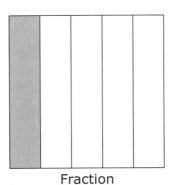
Fraction

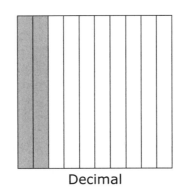
Decimal

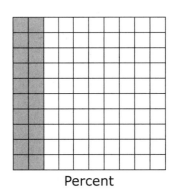
Percent

What values do the models represent? _____ = _____ = _____

Explain how the models are similar and how they are different.

_____
_____
_____
_____
_____

2. Marlene has one dozen cookies she brings to school to share with her friends. Cara also brings cookies to share, but she brings 18 cookies. Marlene takes home $\frac{1}{3}$ of her cookies, and Cara takes home $33\frac{1}{3}\%$ of her cookies. Did the girls take home the same number of cookies? Explain your answer.

_____
_____
_____

How many cookies did each girl take home?

Marlene _____   Cara _____

Unit 20 Journal/Vocabulary Activity

Name _____

Standard 6.5(C) – Supporting

Journal

Explain why calculating 50% does not always give the same result. Use an example to aid in your explanation.

_____
_____
_____
_____

Vocabulary Activity

Draw a line to match each term to its definition. Then, draw a line from the definition to the best example for that term.

Term	Definition	Example
Decimal number	A number written as a ratio, showing the number of parts in a whole	80%
Denominator	A number written using a point to separate the part of the number greater than 1 from the part of the number less than 1	$\frac{7}{9}$
Equivalent	Gives the number of parts of the whole that are being used	$\frac{10}{30}$
Fraction	Gives a ratio per 100	$\frac{1}{2} = 0.5 = 50\%$
Numerator	Having the same value	$\frac{10}{30}$
Percent	Gives the number of parts a whole is divided into	4.18

164   ILLEGAL TO COPY   motivationmath™ LEVEL 6   ©2014 mentoringminds.com

Name _____

Standard 6.5(C) – Supporting

Unit 20 Motivation Station

Creative Mosaic

Create a mosaic (picture) on the grid below. Use the following guidelines to create your mosaic.

You must use at least three colors, but no more than five colors.

All 100 squares must be colored.

Once your mosaic is complete, fill in the chart at the bottom of the page. Calculate the part of the whole picture represented by each color. Then write the fraction, decimal, and percent for that color in the chart.

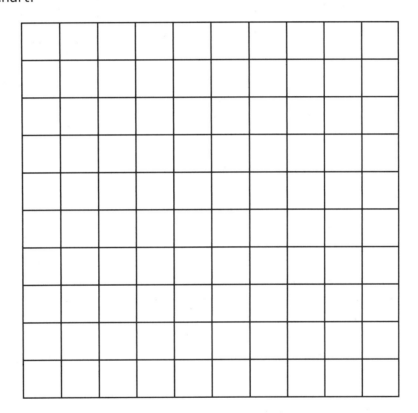

Color	Number of Squares	Fraction	Decimal	Percent

Unit 20 Homework

Standard 6.5(C) – Supporting

1 Use the model below to represent 85%.

0%   100%

What is the fraction equivalent of 85%?

_____

Explain how the model can be used to help answer this question.

_____

_____

_____

2 Molly and Harrison both have Mrs. Long for math. On the last quiz, there were 20 questions. Molly correctly answered 16 questions, and Harrison correctly answered 80% of the questions. Who made a better grade on the quiz? Justify your answer.

_____

_____

_____

3 Destiny, Robin, and Kiara are shopping at the mall on Saturday. Destiny purchases a shirt for $\frac{1}{4}$ off the original price of $24. Robin finds a pair of pants that are 25% off the original price of $49. Kiara buys a skirt that is 0.2 of its original price of $49. Did any of the girls pay the same amount for the item they purchased? Explain.

_____

_____

_____

Use the following information to answer questions 4 and 5.

Mrs. Farmer teaches math. Every year she gives her students a breakdown of how their grades will be calculated in her class. This year's grades will be calculated as follows:

- $\frac{1}{4}$ tests
- 12.5% homework assignments
- $\frac{3}{8}$ classwork and participation
- 0.25 quizzes

4 Which two categories count the same in the grade calculations?

_____

5 What percent of a student's grade depends on classwork and participation?

_____

6 Write the decimal and percent equivalents of $\frac{1}{4}$.

$\frac{1}{4}$ = _____ = _____

Perform each of the calculations below.

$\frac{1}{4} \cdot 16 =$ _____

$$\begin{array}{r} 0.25 \\ \times 28 \\ \hline \end{array}$$

25% of 20 is _____

Did you get the same answer on all problems? Explain. _____

_____

_____

Connections

1. Search the newspaper for ads offering percentage discounts. Find at least two ads with the same discount but different original prices. Attach the ads to a sheet of paper, and explain how the sale prices are similar and how they are different.

2. Write a short paragraph about what you have learned in this unit. It could be a story, a narrative, or a persuasion paper. Share your work with your family.

Standards 6.6(A) – Supporting, 6.6(B) – Supporting

Unit 21 Introduction

Use the information to answer questions 1–3.

Nicky completes chores, and his younger brother Alex helps. Their parents pay them for completed chores. They pay Alex first and then pay Nicky who always earns $2 more than Alex. The graph below shows the relationship between the amounts Alex and Nicky earn.

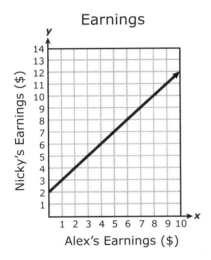

Earnings

1. The amount Nicky earns is dependent on _____.

2. Based on the information above, are Alex's earnings the independent quantity or the dependent quantity?

   _____

   Based on the information above, are Nicky's earnings the independent quantity or the dependent quantity?

   _____

   Explain your answer.

   _____
   _____

3. Fill in the blanks.

   The independent quantities are graphed along the _____-axis.

   The dependent quantities are graphed along the _____-axis.

Use the table to answer questions 4–6.

The table shows a relationship between x and y.

x	-15	-20	-25	-30	-35
y	3	4	5	6	7

4. List the independent quantities shown in the table.

   _____

5. Which numbers in the table represent dependent quantities?

   _____

6. Write an equation to show the relationship between x and y.

   _____

Use the table to answer questions 7–9.

The table shows the costs for student admission to Times Square Theater.

Student Movie Tickets

Number of Tickets (t)	Cost (c)
1	$5.75
2	$11.50
3	$17.25
4	$23.00

7. What label best describes the dependent quantities shown in the table?

   _____

8. What label best describes the independent quantities shown in the table?

   _____

9. Write an equation to describe the relationship between c and t.

   _____

Unit 21 Guided Practice

Standards 6.6(A) – Supporting, 6.6(B) – Supporting

1 Which table shows an independent quantity of 3 with a corresponding dependent quantity of 6?

Ⓐ
x	y
2	4
3	5
4	6
5	7

Ⓒ
x	y
2	4
3	6
4	8
5	10

Ⓑ
x	y
10	5
8	4
6	3
4	2

Ⓓ
x	y
7	6
6	5
5	4
4	3

2 Which of the following could be used to describe the dependent and independent quantities in the graph?

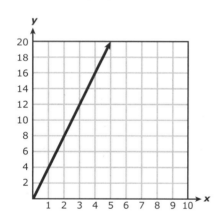

Ⓕ Dependent quantity: number of pints
Independent quantity: number of cups

Ⓖ Dependent quantity: number of quarts
Independent quantity: number of gallons

Ⓗ Dependent quantity: number of cups
Independent quantity: number of pints

Ⓙ Dependent quantity: number of gallons
Independent quantity: number of quarts

3 The graph below shows the relationship between the independent and dependent quantities.

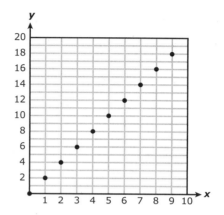

Which set represents all the independent quantities shown in the graph?

Ⓐ {1, 2, 3, 4, 5, 6, 7, 8, 9}

Ⓑ {0, 1, 2, 3, 4, 5, 6, 7, 8, 9}

Ⓒ {2, 4, 6, 8, 10, 12, 14, 16, 18}

Ⓓ {0, 2, 4, 6, 8, 10, 12, 14, 16, 18}

4 The Chic Boutique has a sale on school spirit clothing. The table shows the regular price, r, and the amount of discount, d.

Spirit Clothing Sale

Regular Price (r)	Amount of Discount (d)
$12.00	$3.60
$15.00	$4.50
$22.00	$6.60
$30.00	$9.00

Which equation best represents the relationship r and d?

Ⓕ $d = \dfrac{r}{0.3}$

Ⓖ $r = \dfrac{0.3}{d}$

Ⓗ $r = 0.3d$

Ⓙ $d = 0.3r$

Standards 6.6(A) – Supporting, 6.6(B) – Supporting

Unit 21 Independent Practice

1 The graph below shows the relationship between *x* and *y*.

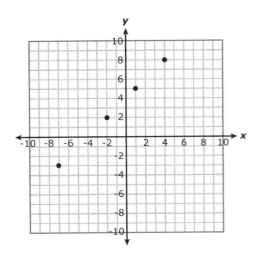

Which set of numbers represents all the dependent quantities shown in the graph?

Ⓐ {-3, 2, 1, 4}　　Ⓒ {-3, 2, 5, 8}

Ⓑ {-7, -2, 1, 4}　　Ⓓ {-3, -2, 5, 8}

2 Samuel uses the equation $y = \frac{2}{5}x$ to create the table of values shown below.

x	1	2	3	4
y	0.4	0.8	1.2	1.8

Which dependent quantity from the table is incorrect?

Record your answer and fill in the bubbles on the grid below. Be sure to use the correct place value.

Use the table to answer questions 3 and 4.

Brianna uses the table to calculate the cost for her babysitting services.

Babysitting

Number of Hours (*h*)	Cost (*c*)
2	$17.00
3	$25.50
4	$34.00
5	$42.50

3 Which equation correctly represents the relationship between the independent and dependent quantities?

Ⓐ $c = h + 8.5$

Ⓑ $h = 8.5c$

Ⓒ $c = 8.5h$

Ⓓ $h = \frac{8.5}{c}$

4 Which of the following is true about the quantities shown in the table?

  I. The values for *c* represent the independent quantities.

  II. The values for *h* represent the dependent quantities.

  III. The values for *c* represent the dependent quantities.

  IV. The values for *h* represent the independent quantities.

Ⓕ I only

Ⓖ I and II only

Ⓗ III only

Ⓙ III and IV only

Unit 21 Assessment

Standards 6.6(A) – Supporting, 6.6(B) – Supporting

Use the graph to answer questions 1 and 2.

The graph shows the cost of single-dip ice cream sundaes at Braum's.

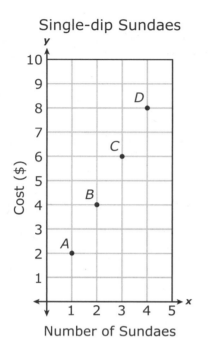

1 Which point on the graph shows 4 as the independent quantity?

- Ⓐ Point A
- Ⓑ Point B
- Ⓒ Point C
- Ⓓ Point D

2 Which statement is true about the set of numbers and its relationship to the graph?

- Ⓕ {1, 2, 3, 4} represents the set of dependent quantities shown in the graph.
- Ⓖ {2, 4, 6, 8} represents the set of independent quantities shown in the graph.
- Ⓗ {1, 2, 3, 4, 6, 8} represents the set of independent quantities shown in the graph.
- Ⓙ {2, 4, 6, 8} represents the set of dependent quantities shown in the graph.

3 Look at the table below.

x	1	2	3	4	5
y	$\frac{1}{4}$	$\frac{1}{2}$	$\frac{3}{4}$	1	$\frac{5}{4}$

Which equation could be used to show the relationship between x and y?

- Ⓐ $x = y + \frac{1}{4}$
- Ⓒ $x = \frac{1}{4}y$
- Ⓑ $y = \frac{1}{4}x$
- Ⓓ $y = x - \frac{1}{4}$

4 Which table shows a dependent quantity of -2 with a corresponding independent quantity of 4?

Ⓕ
x	y
-2	4
-3	3
-4	2
-5	1

Ⓗ
x	y
5	1
4	0
3	-1
2	-2

Ⓖ
x	y
10	-5
8	-4
6	-3
4	-2

Ⓙ
x	y
0	0
-1	4
-2	8
-3	12

5 The table shows the relationship between Santiago's age, s, and Veronica's age, v.

Santiago's Age (s)	Veronica's Age (v)
4	8
8	12
12	16
14	18

Which equation best represents the relationship between the independent and dependent quantities?

- Ⓐ $v = s + 4$
- Ⓒ $v = 2s$
- Ⓑ $s = v + 4$
- Ⓓ $s = \frac{v}{2}$

Name _____

Standards 6.6(A) – Supporting, 6.6(B) – Supporting

Unit 21 Critical Thinking

1. In Mr. Taylor's math class, students built ramps to race toy cars. They tested the hypothesis that the steepness of the ramp affects the distance (in feet) the car travels. Each group in the class was assigned a different height (in inches) to build their ramp. The data collected during the tests is shown in the graph below. Based on the given information, label the *x*- and *y*-axes.

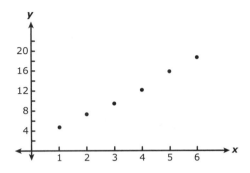

From your answer above, complete the following sentence:

The _____ is dependent on the _____ which is an independent quantity.

2. Jennifer babysits on weekends to earn spending money. She charges $5 per hour for each child. Create a table below to show the amount of money Jennifer earns for one child in relation to the number of hours she works. Then graph the data, labeling the axes, and write an equation to show the relationship.

Hours Worked	Amount Earned
1	
3	
5	
7	
9	

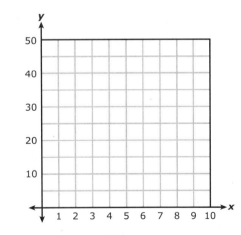

Equation: _____

If Jennifer increases her rate to $6.50 per hour, which quantities will be affected? Discuss how the change will be reflected in the different representations above.

_____

_____

Unit 21 Journal/Vocabulary

Name _____

Standards 6.6(A) – Supporting, 6.6(B) – Supporting

Journal

Explain how to determine whether a set of values in a table is independent or dependent.

_____
_____

Explain how to determine whether a set of values on a graph is independent or dependent.

_____
_____

Vocabulary Activity

A cinquain poem has five lines and does not have to rhyme. Write two cinquain poems using *independent* and *dependent*. For this activity, use the following pattern:

Line 1: One word (given)

Line 2: Two words

Line 3: Three words

Line 4: Four words

Line 5: One word

Independent

_____ _____

_____ _____ _____

_____ _____ _____ _____

_____

Dependent

_____ _____

_____ _____ _____

_____ _____ _____ _____

_____

Name _____

Standards 6.6(A) – Supporting, 6.6(B) – Supporting

Unit 21 Motivation Station

Dependent on What?

Play *Dependent on What?* with a partner. Each pair of players needs a game board, one number cube, and a paper clip to use with the spinner. Each player needs a pencil. Player 1 rolls the number cube and spins the spinner. The player matches the number rolled to the table with the same number. He or she then completes the section in the table at the bottom of the page that matches the space on the spinner, using the information from the numbered table. If answered correctly, the player initials the space and play passes to the next player. If incorrect, the player loses a turn and play passes to the next player. If the section has already been initialed, the player spins and rolls again. The game ends when all sections of the table are complete. The winner is the player with more sections initialed.

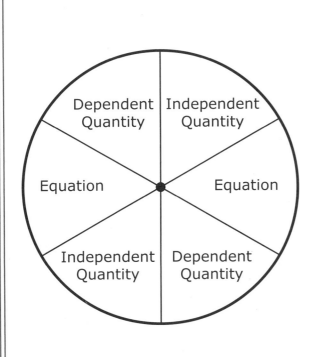

1.
x	-1	0	1	2
y	3	0	-3	-6

2.
Cups of Sugar	$\frac{3}{4}$	$1\frac{1}{2}$	$2\frac{1}{4}$	3
Dozen Cookies	3	6	9	12

3.
Miles Driven	116.5	163.1	209.7	256.3
Gallons Used	5	7	9	11

4.
Number of Weeks	1	2	3	4
Growth (inches)	0.5	1	1.5	2

5.
Number of Slices	2	4	6	8
Total Cost	$3.50	$7	$10.50	$14

6.
a	-1	0	1	2
b	$-\frac{2}{3}$	0	$\frac{2}{3}$	$1\frac{1}{3}$

	1	2	3	4	5	6
Independent Quantity						
Dependent Quantity						
Equation						

Unit 21 Homework

Name _____

Standards 6.6(A) – Supporting, 6.6(B) – Supporting

Use the information and table to answer questions 1–3.

Walter has a lawn mowing business. He charges $25 for each lawn he mows. Complete the table below.

Number of Lawns Mowed, m	1	2	3	4	5
Amount of Money Earned, d					

1. List the independent quantities from the table.

   _____

2. List the dependent quantities from the table.

   _____

3. Write an equation to represent the data in the table.

   _____

4. The table below shows the relationship between regular prices and sale prices in a clothing shop.

   ### Clothing Shop Prices

Regular Price (r)	Sale Price (s)
$18	$12
$24	$16
$27	$18
$36	$24

   Write an equation to represent the sale price given the regular price.

   _____

5. For the day's math lesson, students in Mr. Carson's class were asked to list their age and shoe size in a chart on the board. Mr. Carson then created the following graph.

   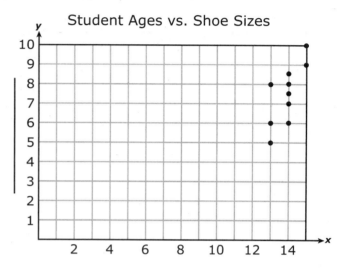

   Student Ages vs. Shoe Sizes

   _____

   What is an appropriate label for the independent quantities on the graph? Write the label in the blank above.

   What is an appropriate label for the dependent quantities on the graph? Write the label in the blank above.

6. Write an equation to represent the relationship between the independent and dependent quantities in the following table.

x	1	2	3	4
y	-3	-2	-1	0

   _____

Connections

1. Interview your science teacher about independent and dependent variables and how they are used in science. Write a paragraph describing the similarities and differences of the terms *independent* and *dependent* as used in math and science.

2. Find two tables and two graphs in a newspaper. Cut and glue the examples on a sheet of paper. Under each example, identify the independent quantities and the dependent quantities.

Name _____

Standard 6.6(C) – Readiness

Unit 22 Introduction

Complete the chart below using the following word problem.

Mr. Wimmer stops at the donut shop every morning to buy a donut and a cup of coffee. He spends $0.75 for the donut and $2.75 for the coffee. Represent the total amount of money Mr. Wimmer spends for different numbers of days.

1. Write a short verbal description of the mathematics in the word problem.

2. Define your variables. Write an equation to show the relationship between x and y.

   x: _____

   y: _____

   Equation: _____

   Does the equation show an additive relationship or a multiplicative relationship?

   _____

3. Use the table below to calculate the dependent quantities from the corresponding independent quantities.

x	y
1	
2	
3	
4	
5	

   How much money does Mr. Wimmer spend at the donut shop in one week? _____

4. Use the coordinate plane to graph the data from the table. Be sure to label the x- and y-axis and give the graph a title.

   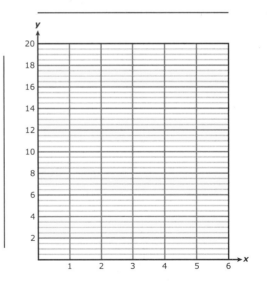

   Should the points on the graph be connected? Explain your reasoning.

Unit 22 Guided Practice

Standard 6.6(C) – Readiness

1. Derrick's favorite game at the fair is basketball. He wants to bring enough money to play the game several times. If each game costs $2.50 to play, which table represents the amount of money Derrick might spend playing basketball at the fair?

Ⓐ
x	y
0	0
1.50	1.50
2.50	2.50
3.50	3.50

Ⓒ
x	y
0	0
1	2.50
2	5.50
5	10.00

Ⓑ
x	y
0	0
2.50	1
4.50	2
6.50	3

Ⓓ
x	y
0	0
1	2.50
4	10.00
7	17.50

2. The graph below shows the relationship between Jace's age and Tyson's age.

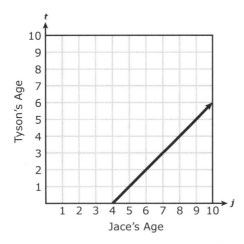

Which description best represents this relationship?

Ⓕ Jace's age is 4 years less than Tyson's.

Ⓖ Tyson's age is 4 years less than Jace's.

Ⓗ Jace's age is 1 year more than Tyson's.

Ⓙ Tyson's age is 4 years more than Jace's.

3. Mya uses the equation $y = \frac{1}{4}x$ to determine the number of gallons, x, for a number of quarts, y. Which table does NOT represent the relationship between quarts and gallons?

Ⓐ
x	3	7	10	15
y	$\frac{3}{4}$	$1\frac{3}{4}$	$2\frac{1}{2}$	$3\frac{3}{4}$

Ⓑ
x	2	4	11	20
y	$\frac{1}{2}$	1	$3\frac{3}{4}$	5

Ⓒ
x	5	8	12	17
y	$1\frac{1}{4}$	2	3	$4\frac{1}{4}$

Ⓓ
x	6	9	13	14
y	$1\frac{1}{2}$	$2\frac{1}{4}$	$3\frac{1}{4}$	$3\frac{1}{2}$

4. Which of the following statements best describes the data shown in the table?

x	1	2	3	4
y	6	7	8	9

Ⓕ The dependent quantity is 5 more than the independent quantity.

Ⓖ The independent quantity is 5 more than the dependent quantity.

Ⓗ The dependent quantity is 5 times as much as the independent quantity.

Ⓙ The independent quantity is 5 times as much as the dependent quantity.

Standard 6.6(C) – Readiness

Unit 22 Independent Practice

1. Mel walks each evening at a constant rate of speed. She walks 2 miles in a half hour. Which graph best represents the relationship between time and distance traveled?

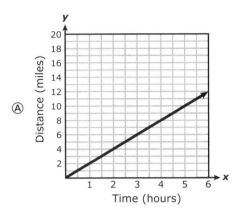

Ⓐ

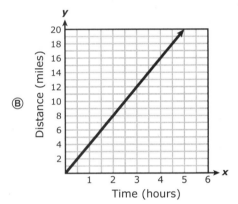

Ⓑ

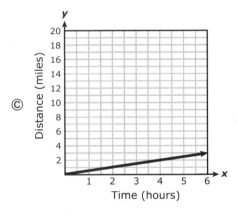

Ⓒ

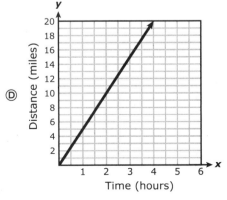

Ⓓ

2. An Internet business charges a flat rate of $6.50 for shipping. If x represents the cost of the items purchased and y represents the total cost with shipping, which equation best represents the relationship between x and y?

Ⓕ $y = 6.5x$ Ⓗ $y = \frac{x}{6.5}$

Ⓖ $y = x + 6.5$ Ⓙ $y = x - 6.5$

3. The table below represents the relationship between x and y in question 2.

Cost of Items (x)	Total Cost with Shipping (y)
$1.00	$7.50
$3.00	$9.50
$5.00	$11.50

Which ordered pair would NOT be contained in the table above?

Ⓐ (2.25, 8.75) Ⓒ (9.5, 15)

Ⓑ (4, 10.5) Ⓓ (12.75, 19.25)

4. Kendra was asked to explain the relationship between x and y in the table.

x	1	$\frac{3}{2}$	$\frac{9}{4}$	$\frac{15}{4}$
y	$\frac{3}{4}$	$\frac{5}{4}$	2	$\frac{7}{2}$

Which best explains the relationship between the x and y values?

Ⓕ The value of y is $\frac{1}{4}$ less than the value of x.

Ⓖ The value of y is $\frac{3}{4}$ times the value of x.

Ⓗ The value of y is $\frac{1}{4}$ greater than the value of x.

Ⓙ The value of y is $\frac{3}{4}$ less than the value of x.

Unit 22 Assessment

Standard 6.6(C) – Readiness

1 The Raibourn family drove to the beach for vacation. The graph below shows the relationship between the time and distance traveled.

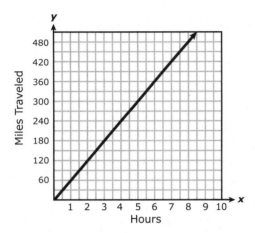

Which table best shows the relationship between the number of hours and the miles traveled as shown in the graph?

Ⓐ
x	2.5	3	6	8.5
y	150	200	360	510

Ⓑ
x	2.5	3	6	8.5
y	150	200	360	510

Ⓒ
x	0.5	3.5	5	8
y	30	200	300	460

Ⓓ
x	4	5.5	6.5	8.5
y	240	330	390	510

2 Juan Carlos is j years old. Sara's age, s, is 9 months older than Juan Carlos. Which equation best describes the relationship, in years, between the ages of Juan Carlos and Sarah?

Ⓕ $s = j + 9$ Ⓗ $s = j - 9$

Ⓖ $j = s - 0.75$ Ⓙ $j = s + 0.75$

3 Which of the following statements does NOT describe the data shown in the table?

x	80	64	40	16
y	10	8	5	2

Ⓐ The value of x is 8 more than the value of y.

Ⓑ The value of y is $\frac{1}{8}$ the value of x.

Ⓒ The value of y is the value of x divided by 8.

Ⓓ The value of x is 8 times the value of y.

4 Felipe pays $4.20 per carton for fresh eggs from the farmers' market. If x represents the number of egg cartons and y is the total cost, which of the following representations contains an error?

Ⓕ $y = 4.2x$

Ⓖ
x	1	3	4
y	4.20	12.60	16.80

Ⓗ The value of y is $\frac{21}{5}$ times the value of x.

Ⓙ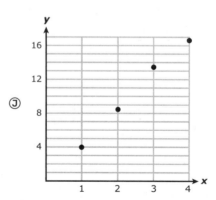

Standard 6.6(C) – Readiness

Unit 22 Critical Thinking

1 April works at the local grocery store after school and on weekends. She earns $7.25 per hour, and she is allowed to work up to 20 hours per week. Three different representations of the amount of money April earns are shown below.

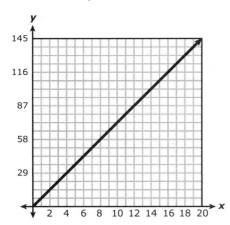

x	y
0	0
4	29
8	58
12	87
16	116
20	145

$y = 7.25x$

Explain how the three representations are related. Include in your explanation the answers to the following questions: Is it possible to create the other two representations when only one representation is given? Does it matter which representation is given first?

_____
_____
_____

2 Write your own word problem using the pattern $y = x + a$ or $y = ax$. Then, using the information from the problem, complete the table and graph and write an equation that represents the situation.

Equation: _____

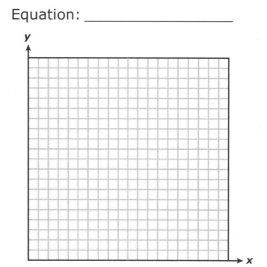

x					
y					

Unit 22 Journal/Vocabulary

Standard 6.6(C) – Readiness

Journal

You have learned multiple ways to represent linear relationships: tables, graphs, equations, and verbal descriptions. Which representation do you think is best? Justify your choice.

_____
_____
_____

Vocabulary Activity

Classify the following representations into a category by labeling each box with the appropriate letter: **A** – algebraic representation, **G** – graphical representation, **V** – verbal description, **T** – tabular representation.

A
$y = \frac{3}{4}x$

B

a	0	1	2	3
b	3	4	5	6

C
For every 16 bales of hay Gerald loads, Timothy loads 14.

D
A representation using numbers, variables, and operational symbols; an equal sign is included in some cases

F
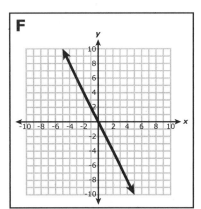

G
A representation created using data points and a coordinate plane

E
$p + 7 = m$

K
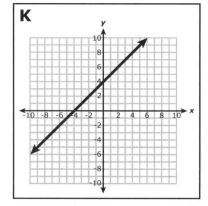

H

r	t
-6	$-2\frac{1}{4}$
-4	$-1\frac{1}{2}$
-2	$-\frac{3}{4}$
0	0
2	$\frac{3}{4}$
4	$1\frac{1}{2}$
6	$2\frac{1}{4}$

J
A representation that organizes data using rows and columns

L
Roy's Plumbing charges $75 per hour for a service call.

M
A representation that describes math in words

Standard 6.6(C) – Readiness

Unit 22 Motivation Station

Equal Representation

Play *Equal Representation* with a partner. Each pair of players needs a number cube, a game board, and a paper clip to use with the spinner. Each player needs a pencil. Player 1 rolls the number cube and spins the spinner. Player 1 uses the data from the situation on the spinner to complete a representation based on the number rolled and the instructions below. Player 1 initials his/her work when complete, and play passes to player 2. If the selection rolled for the situation has already been completed, the player loses a turn. The game ends when all four representations are complete for all three situations. The winner is the player with more initials on the board.

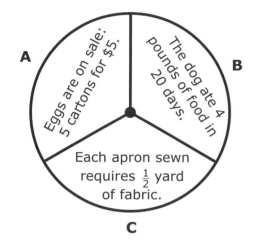

Number Rolled
1: Complete the table.
2: Draw the graph.
3: Write the equation.
4: Write a verbal description of the math in the situation.
5: Player chooses which representation to complete.
6: Partner chooses which representation to complete.

A

x	0	1	2	3	4
y					

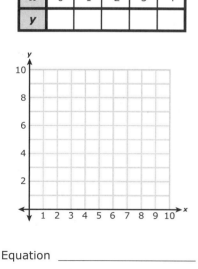

Equation _____

Verbal _____
_____

B

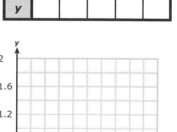

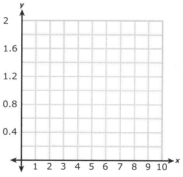

Equation _____

Verbal _____
_____

C

x	0	1	2	3	4
y					

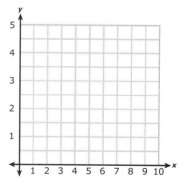

Equation _____

Verbal _____
_____

Unit 22 Homework

Name _____

Standard 6.6(C) – Readiness

Match each situation below with its correct representation on the right by writing the correct letter in the blank. Then, answer the question.

1. _____ For every cookie Jessica places in a gift box, she adds 6 pieces of candy. How many pieces of candy are in a box with 6 cookies?

   _____

2. _____ To rent a party room, customers are charged an amount equal to the number of hours they rent plus a $60 cleaning fee. What is the cost of a 3-hour rental?

   _____

3. _____ Three bags of topsoil weigh a total of 120 pounds. How much do 10 bags weigh?

   _____

4. _____ Becca and Marian run at the same pace. Becca begins running first. When she has run 6 laps, Marian has run 3 laps. How many laps has Becca run if Marian runs 7 laps?

   _____

5. _____ Claudia worked 18 hours at her job last week. She earned $135. How much will Claudia's check be if she works 22 hours next week?

   _____

6. _____ Jeffrey mixes 2 cups of water for every $\frac{1}{2}$ cup of orange concentrate. How much water will he use for 4 cups of concentrate?

   _____

A. $y = 4x$

B.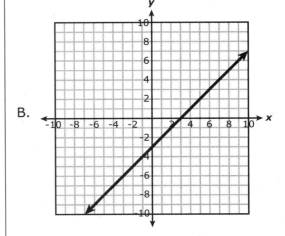

C.
x	0	1	2	3	4
y	0	6	12	18	24

D. For every one y, there are 7.5x.

E. $y = 40x$

F.
x	1	3	5	7	9
y	61	63	65	67	69

Connections

1. Look for a table, graph, or equation in a newspaper or magazine. Glue the representation on a sheet of paper and create the other 2 representations. Present the work to the class.

2. Explore other areas outside of math where different representations are used to model the same thing. Make a list of your findings to share and discuss with the class.

Standard 6.7(A) — Readiness

Unit 23 Introduction

1 Mr. Brandt works with a small group of students. He gives each student a different expression to evaluate as listed below.

Tabitha: $17 + 4(5^2 - 10)$

Martin: $\dfrac{(7+2)^2}{3} \cdot 3$

Lucy: $56 + (34 + 4^2) \div 2$

Everly: $\dfrac{18}{3} + 5 \cdot 2^4$

Which students have equivalent expressions?

_____

2 Simplify the following expression.

$10 - 7.2 + 0.4 \cdot 5^2 \div 10$

3 Create a factor tree for the number 84.

Write an expression using prime factorization that is equivalent to 84.

4 Mrs. Maguire requires each student to correct questions missed on a test. Lexi incorrectly simplified this expression on her last test.

Expression: $26 + (10 - 6^2) \cdot \dfrac{5}{2}$

Line 1: $26 + (4^2) \cdot \dfrac{5}{2}$

Line 2: $26 + 16 \cdot \dfrac{5}{2}$

Line 3: $26 + 40$

Line 4: 66

On which line did Lexi make her first mistake?

_____

Show the correct steps to simplify the expression.

5 Place a set of parentheses in the expression below so that the expression is equivalent to 70. Then show the steps to prove your answer is correct.

$28 + 24 \div 8 - 2^2 \cdot 7$

6 Sam and Layla work on their homework together after school. They disagree about the prime factorization of 324.

Sam: $324 = 2^2 \cdot 3^4$

Layla: $324 = 2^2 \cdot 9^2$

Who is correct? _____
Explain your answer.

_____

Unit 23 Guided Practice

Standard 6.7(A) – Readiness

1. Mrs. Wong found a mistake as she was grading Lola's paper. Each step should represent an equivalent expression. Which step shows the mistake that Mrs. Wong found?

   Expression: $\frac{3}{2}(8^2 + 4) - 3\frac{1}{2}$

   Step 1: $\frac{3}{2}(64 + 4) - 3\frac{1}{2}$

   Step 2: $\frac{3}{2} \cdot 68 - 3\frac{1}{2}$

   Step 3: $\frac{3}{2} \cdot 64\frac{1}{2}$

   Step 4: $96\frac{3}{4}$

   Ⓐ Step 1   Ⓒ Step 3
   Ⓑ Step 2   Ⓓ Step 4

2. Lizbeth evaluated the following expression.

   $20 + (6 \cdot 5) + 10 - 3^2$

   By moving the parentheses, Lizbeth noticed she could create an equivalent expression. Which best represents the expression Lizbeth might have written?

   Ⓕ $(20 + 6) \cdot 5 + 10 - 3^2$

   Ⓖ $20 + 6 \cdot 5 + (10 - 3^2)$

   Ⓗ $20 + 6 \cdot (5 + 10) - 3^2$

   Ⓙ $20 + 6 \cdot (5 + 10 - 3^2)$

3. Diego found the prime factorization for each number below.

   $90 = 2 \cdot 3^2 \cdot 5$ $250 = 2 \cdot 5^3$
   $400 = 2^3 \cdot 5^2$ $110 = 2 \cdot 5 \cdot 11$

   Which prime factorization does NOT correctly represent the number shown?

   Ⓐ 90   Ⓒ 250
   Ⓑ 400   Ⓓ 110

4. What is the value of the expression below?

   $(25 - 6.25) + (5 \cdot 3)^2$

   Record your answer and fill in the bubbles on the grid below. Be sure to use the correct place value.

5. The expression below is equivalent to which integer?

   $2(\text{-}57 \div 3) + 20^2 - 5$

   Ⓐ 357
   Ⓑ 433
   Ⓒ -3
   Ⓓ Not here

6. Mr. Lummus said that the prime factorization of 540 could be represented by the expression below.

   $x^2 \cdot y^3 \cdot z$

   What values of x, y, and z are true for the expression?

   Ⓕ $x = 2, y = 3, z = 4$
   Ⓖ $x = 3, y = 2, z = 5$
   Ⓗ $x = 2, y = 3, z = 5$
   Ⓙ $x = 2, y = 3, z = 7$

Name _____

Standard 6.7(A) – Readiness

Unit 23 Independent Practice

1. As Thomas was checking his homework, he found a mistake in simplifying the expression below. Each step in the process should represent an expression equivalent to the original problem. Which step shows the error that Thomas found?

    Simplify: $7 - 2.4 + 3(0.8^2 \div 2)$

    Step 1: $7 - 2.4 + 3(0.16 \div 2)$

    Step 2: $7 - 2.4 + 3(0.08)$

    Step 3: $7 - 2.4 + 0.24$

    Step 4: $4.6 + 0.24$

    Step 5: 4.84

    Ⓐ Step 1

    Ⓑ Step 2

    Ⓒ Step 3

    Ⓓ Step 4

2. What is the value of the expression below?

    $(-8 + 3)^2 + 10 \cdot (-4)$

    Record your answer and fill in the bubbles on the grid below. Be sure to use the correct place value.

3. Mrs. Morrison asked the class to generate an expression equivalent to 80. Lucy used the prime factorization of 80 to form an equivalent expression. Which best represents the expression Lucy wrote?

    Ⓐ $10 \cdot (2 + 6)$

    Ⓑ $2^4 \cdot 5$

    Ⓒ $(5^2 - 5) \cdot 4$

    Ⓓ $2 \cdot 5 \cdot 8$

4. Complete the factor tree below.

    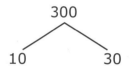

    Which expression shows the correct prime factorization of 300?

    Ⓕ $10 \cdot 30$ Ⓗ $2^2 \cdot 3 \cdot 5^2$

    Ⓖ $2^3 \cdot 3 \cdot 25$ Ⓙ $2^3 \cdot 3 \cdot 5^2$

5. Which of the following expressions are equivalent to 30?

    I. $3^2 + 7(12 \div 4)$

    II. $25 - 7 + 2^3 \cdot \frac{3}{2}$

    III. $3(20 - 4^2) \cdot 2 + 6$

    IV. $\frac{7^2 - 9}{5} + 11 \cdot 2$

    Ⓐ I and III only   Ⓒ I, III, and IV only

    Ⓑ II and IV only   Ⓓ I, II, III, and IV

Unit 23 Assessment

Name _____

Standard 6.7(A) – Readiness

1 Simplify the expression below.

$$\frac{1}{2} \cdot 6 + 9^2 - (18 \div 6)$$

Which shows the value of the simplified expression?

Ⓐ 118

Ⓑ 81

Ⓒ 18

Ⓓ 42

2 Max needs to create equivalent expressions that simplify to 12. Which of Max's expressions is NOT equivalent to 12?

Ⓕ $6^2 \div 4 + 3 \cdot 1$

Ⓖ $(2^3 - 6) \cdot \frac{12}{2}$

Ⓗ $(9^2 - 9) \div 8 + 4 - 2$

Ⓙ $2^2 + 2^3$

3 Hayli and Jacqueline are instructed to create two equivalent expressions using integers. Which pair of expressions could Hayli and Jacqueline create?

Expression 1: $5(-8) + 6^2 + (-24 \div 4)$

Expression 2: $-10 + 2^2(-3 - 2)$

Expression 3: $2^4 + (-6 + 20) \cdot \frac{-12}{4}$

Expression 4: $\left(\frac{36}{9}\right)^2 - 13 \cdot 2$

Ⓐ Expressions 1 and 2

Ⓑ Expressions 2 and 3

Ⓒ Expressions 3 and 4

Ⓓ Expressions 1 and 4

4 Blake wrote the following expression for the prime factorization of 513:

$$3^2 \cdot 57$$

Which statement is true about the expression Blake wrote?

Ⓕ Blake's expression is incorrect because it is not equivalent to 513.

Ⓖ Blake's expression is correct because both 3 and 57 are prime numbers.

Ⓗ Blake's expression is incorrect because 57 is not a prime number.

Ⓙ Blake's expression is correct because it is equivalent to 513.

5 What is the value of the expression below?

$$6^2 + 3.5 \cdot 3^3 - 7$$

Record your answer and fill in the bubbles on the grid below. Be sure to use the correct place value.

Name _____

Standard 6.7(A) – Readiness

Unit 23 Critical Thinking

1. Carrie's mother bought 5 pounds of mushrooms for $0.79 per pound, 0.5 pound of poblano peppers for $1.38 per pound, and a 5-pound sack of potatoes for $3. She paid for her groceries with a $20 bill. Carrie writes the following equation to find the amount of change her mother received when she paid for her groceries:

$$c = 20 - 5 \cdot 0.79 + 0.5 \cdot 1.38 + 3$$

Did Carrie write the equation correctly? _____

Explain. _____

_____

How much change did Carrie's mother receive? Show all your work below, beginning with the equation.

2. Create an expression that is equivalent to each of the following numbers. Use the following guidelines in writing your expression:

- The only number that may be used is 3.
- Use at least two different operations in each expression.
- You may use parentheses and/or exponents, but the exponent used must be 3.

Expression 1: equivalent to 100

Expression 2: equivalent to 24

Unit 23 Journal/Vocabulary

Name _____

Standard 6.7(A) – Readiness

Journal

What would happen if people did not use sequencing and order in their lives? Write about why you think doing things in a certain order is sometimes necessary and what you think might happen if we stopped using sequencing and order in our lives.

_____
_____
_____

Vocabulary Activity

Complete each graphic organizer below.

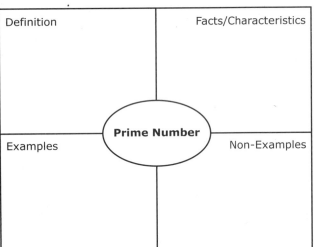

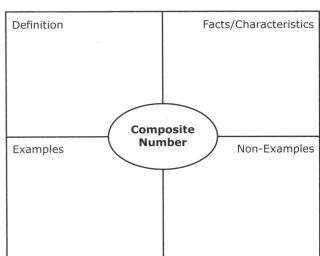

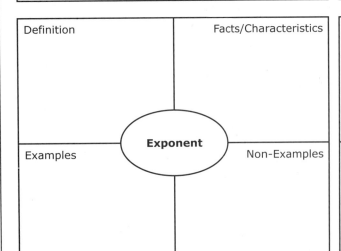

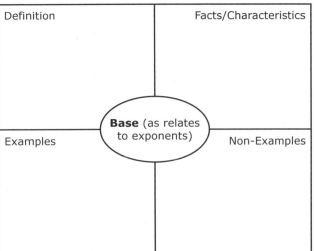

Name _____

Standard 6.7(A) – Readiness

Unit 23 Motivation Station

Order, Order

Play *Order, Order* with a partner. Each pair of players needs 6 number cubes, a calculator for checking answers, and a game board. Each player needs a pencil. Player 1 rolls the number cubes and uses the numbers rolled to fill in the blanks. The numbers may be used in any order. Player 1 then simplifies the expression created. If simplified correctly, player 1 initials the expression, and play passes to player 2. If not simplified correctly, player 2 may steal the expression by correctly solving it. The game ends when all ten expressions have been simplified. The player with more expressions initialed wins.

(___ + ___) · ___ − ___² + ___ · ___	___ · ___ + ___ + ___ · ___ + ___
(___ + ___) · (___ + ___) − ___ + ___²	___ · ___ · ___ ÷ ___ + □/□
(___ · ___ · ___ − ___) ÷ ___ + ___⁴	___ · (□/□ + □/□) · ___
(___ − ___ ÷ ___ − ___²) × □/□	___ − ___ ÷ ___ + ___ · ___ − ___
(___ + ___ − ___)² · □/□ + ___	___ − (___ + ___ · ___) + ___ · ___

©2014 mentoringminds.com   motivationmath™ LEVEL 6   ILLEGAL TO COPY   189

Unit 23 Homework

Name _____

Standard 6.7(A) – Readiness

1 Using the following expression, $9 + 6 \div 3 - 2 \cdot 4$, insert two sets of parentheses so the value of the expression is equivalent to 60.

2 Write the prime factorization of 96 using exponents. Show all work.

3 Danica simplified the following expression.

$$20 - 2(4 + 1) + 3^2$$
$$= 20 - 2(5) + 3^2$$
$$= 20 - 2(5) + 9$$
$$= 20 - 10 + 9$$
$$= 20 - 19$$
$$= 1$$

Circle Danica's mistake. Correctly simplify the expression, showing your work below.

4 Kendrick wrote the following prime factorization of a number.

$$7^2 \times 13^2$$

What number did Kendrick factor? Show all work.

5 Which of the following expressions have integer solutions? Circle your response and write the correct solution in the blank.

$$12 \div \tfrac{3}{2} - 8\left(\tfrac{5}{2}\right)$$

Yes/No   Solution _____

$$-\tfrac{3}{2} \cdot \left(-\tfrac{4}{9}\right) \div \left(-\tfrac{8}{15}\right)$$

Yes/No   Solution _____

$$-8 + 1.2 \div (-0.3) \cdot 1.5$$

Yes/No   Solution _____

$$(2 + 6) \div 2 \div 8$$

Yes/No   Solution _____

Connections

1. Research the origin of the order of operations. Are there alternative methods used in other countries that yield the same results? Write a one-page paper on your findings to share with the class.
2. Create a short presentation either for or against having order in the world, including in mathematics. Present your argument to the class and defend your position.

Name _____

Standards 6.7(B) – Supporting, 6.7(C) – Supporting

Unit 24 Introduction

1 Determine whether each is an expression or an equation.

2 + 12 _____

The distance Keatyn ran is two more miles than the distance Rylan ran. _____

$8s - 10 = -3$ _____

Twice as many cookies _____

$\frac{c}{5}$ _____

2 Explain the similarities and differences between expressions and equations.

_____
_____
_____
_____

3 Isaiah and Ron are working together on their math homework. Isaiah says that $3(x + 3)$ is equivalent to $3x + 3$. Ron says that the two expressions are not equivalent. Using Algebra Tiles™, sketch each expression to show who is correct.

Who is correct? Explain.

_____
_____
_____

4 A bag of fruit contains 5 apples and 4 oranges. The expression $5b + 4b$ represents the total number of pieces of fruit in b bags. Write an expression that is equivalent to $5b + 4b$.

_____
_____

5 Explain the difference between a numerical equation and an algebraic equation. Give an example of each.

_____
_____
_____
_____
_____

6 Look at the following expressions.

$5(3) + 10$

$5(3 + 10)$

$3 + 3 + 3(3) + (2 \times 5)$

Are the expressions equivalent? _____

Justify your answer using words, numbers, or pictures.

Unit 24 Guided Practice

Standards 6.7(B) – Supporting, 6.7(C) – Supporting

1 Mrs. Jacobson asks each group of students to write an expression equivalent to the model below.

x	1	x	1	x	1
x	1	x	1	x	1
x	1	x	1	x	1
x		x		x	

The table shows the groups' expressions.

Group	Expression
1	$3(4x) + 3(3)$
2	$4x(3 + 3 + 3)$
3	$3(4x + 3)$
4	$4x + 3 + 4x + 3 + 4x + 3$

Which group wrote an incorrect expression?

Ⓐ Group 1
Ⓑ Group 2
Ⓒ Group 3
Ⓓ Group 4

2 Which of the following is true for $2g + 6$?

  I. Representative of an equation
  II. Multiple possible values for g
  III. Representative of an expression
  IV. Only one possible value for g

Ⓕ I only
Ⓖ III and IV only
Ⓗ II and III only
Ⓙ I, II, and III only

3 Nia plans to give one necklace and one sparkle pen to each girl who attends her birthday party. Necklaces cost $4 each and sparkle pens cost $2 each. She uses the expression $4g + 2g$ to calculate the cost, in dollars, of the gifts for the girls attending the party. Which of the following is equivalent to Nia's expression?

Ⓐ $6(g + g)$
Ⓑ $6g$
Ⓒ $6g^2$
Ⓓ $8g$

4 Meghan tries to convince her classmate Fiona that $2x + 3 = 10$ represents an expression. Fiona disagrees and thinks that it represents an equation. Who is correct?

Ⓕ Meghan is correct because expressions contain variables.
Ⓖ Fiona is correct because equations always contain an equal sign.
Ⓗ Both girls are correct because expressions and equations are the same thing.
Ⓙ Neither girl is correct.

5 Kareem and Natalie each created an expression using the models shown below.

Kareem's Expression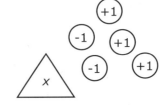

Natalie's Expression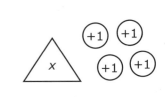

What would Natalie need to add to her model for it to be equivalent to Kareem's model?

Ⓐ (+1)
Ⓑ (-1) (-1)
Ⓒ (-1) (-1) (-1)
Ⓓ (+1) (+1) (+1)

6 Which of the following does NOT represent an equation?

Ⓕ Twice a number is six.
Ⓖ $x + 6 = 10$
Ⓗ 6 years older than Mike
Ⓙ $-9 + 2 = -7$

Standards 6.7(B) – Supporting, 6.7(C) – Supporting

Unit 24 Independent Practice

1 Which of the following is NOT true about equations?

Ⓐ Must include an equal sign

Ⓑ A phrase that contains a single term

Ⓒ Show two equivalent expressions

Ⓓ May contain numbers and variables

2 Which of the following could NOT be used to represent an expression equivalent to the model shown below?

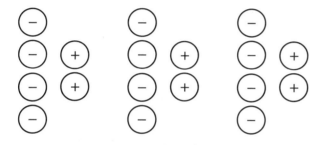

Ⓕ $3(-4) + 3(2)$

Ⓖ $-12 + 6$

Ⓗ $-4(3 + 2)$

Ⓙ $3(-2)$

3 Which of the following represents an expression?

   I. Four years less than Tiffany's age

   II. The sum of a number and six

   III. Tim saves $30 more than Jeremy

   IV. The difference of seven and three

Ⓐ II only

Ⓑ I and II only

Ⓒ III only

Ⓓ I, II, and IV only

4 Look at the model below.

1	1	x
1	1	x
1	1	x

Which of the following statements of equality are true?

Ⓕ $3(x + 2) = 3x + 5$

Ⓖ $3x + 26 = 2(3 + 2x)$

Ⓗ $3(2 + x) = 6 + 3x$

Ⓙ $6(2 + 2x) = 12 + 2x$

5 Which of the following best represents an expression?

Ⓐ $-2x + 20$ Ⓒ $-3y = 27$

Ⓑ $13 + 12 = 25$ Ⓓ $\frac{f}{5} = 30$

6 Kelvin modeled an expression using the Algebra Tiles™ shown.

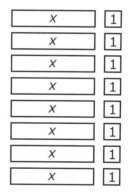

Which of the following could NOT be the expression Kelvin modeled?

Ⓕ $2(4x + 4)$

Ⓖ $6 + 4x + 2x + 2 + x + x$

Ⓗ $4(2x) + 4(2)$

Ⓙ $3x + 2 + 3x + 2 + 3x + 4$

Unit 24 Assessment

Standards 6.7(B) – Supporting, 6.7(C) – Supporting

1 Which model is equivalent to $3(2x + 3)$?

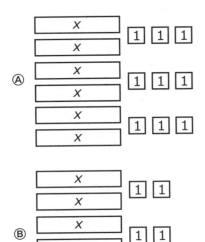

2 Homestyle Kitchens sells precooked holiday meals. A complete meal includes a smoked ham, vegetables, and desserts for a family of four. Smoked hams cost $55 per family, vegetables cost $50 per family, and desserts cost $40 per family. The manager uses the expression $55f + 50f + 40f$ to calculate the amount of money the business will receive for selling f complete family meals. Which of these is NOT equivalent to the manager's expression?

Ⓕ $55f + 90f$

Ⓖ $105f + 40f$

Ⓗ $f + 145$

Ⓙ $145f$

3 Greg wrote in his math journal that $17 - x = 9$ is an example of an algebraic equation. Which statement is true about the example Greg wrote in his math journal?

Ⓐ Greg's example is an equation because there is a variable present.

Ⓑ Greg's example is an equation because it is a number sentence stating two amounts are equivalent.

Ⓒ Greg's example is an equation because x can represent any value.

Ⓓ Greg's example is an equation because there are two numbers present.

4 Which of the following represents an expression?

  I. Three points more than Roman

  II. Twice as many jelly beans as Cynthia

  III. Lydia ran 1.5 miles more than Wally

  IV. Mikki is four inches shorter than Lucas

Ⓕ I only

Ⓖ IV only

Ⓗ I and II only

Ⓙ I, II, and III only

5 Which of the following is NOT true for $6s + 2s + 5$?

Ⓐ Represents an algebraic expression

Ⓑ There is only one value for s

Ⓒ A phrase that simplifies to two terms

Ⓓ There is a solution for $s = 5$

Name _____

Standards 6.7(B) – Supporting, 6.7(C) – Supporting

Unit 24 Critical Thinking

1 The athletic booster club at a middle school is hosting a school dance to raise money for new basketballs. Three students sold tickets prior to the dance.

- Haley sold twice as many tickets as Wesley.
- Sam sold 10 more tickets than Haley.

If w represents the number of tickets sold by Wesley, write two equivalent expressions to show the number of tickets the three students sold together.

_____

_____

Explain how you know your expressions are equivalent. Use math vocabulary and models in your explanation.

2 There were 295 tickets sold to the dance. Using the information in the problem above, write an equation that could be used to determine the number of tickets each student sold.

Use the Venn diagram below to show the similarities and differences between the expressions in question 1 and the equation in question 2.

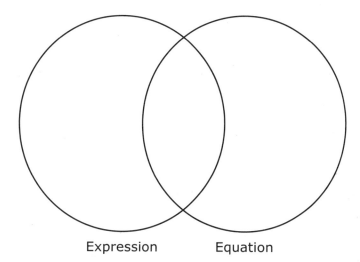

Expression   Equation

Unit 24 Journal/Vocabulary

Standards 6.7(B) – Supporting, 6.7(C) – Supporting

Journal

The teacher leaves the room and you are in charge of continuing the lesson. How do you explain the difference between an algebraic expression and an equation to your classmates?

Vocabulary Activity

Use the clues below to complete the crossword puzzle.

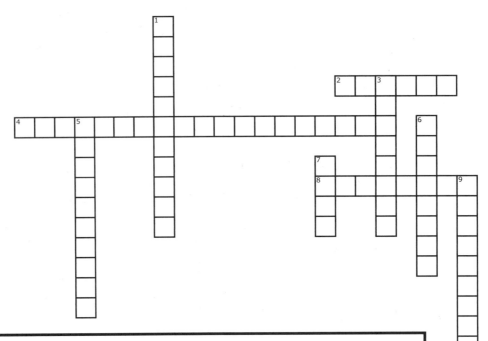

Across

2. a number multiplied with another number or variable
4. a combination of numbers, variables, and operator symbols (two words)
8. two expressions shown to be equivalent with an equal symbol

Down

1. numeric factor accompanying a variable, such as $4x$
3. a numeric term in an expression or equation, such as the 4 in $x + 4$
5. a group of terms separated by operator symbols
6. symbol that represents a quantity in an algebraic expression or equation
7. either a single number or variable, or numbers and variables multiplied together
9. an expression containing only numbers and operator symbols (two words)

Name _____

Standards 6.7(B) – Supporting, 6.7(C) – Supporting

Unit 24 Motivation Station

Matchmaker

Play *Matchmaker* with a partner. Each pair of players needs 16 Color Tiles™, 8 each of two colors, and a game board. Place tiles on the game board squares, one color covering the expressions and the second color covering the models. Player 1 takes away two tiles, one of each color. If the expression revealed and the model revealed are equivalent, player 1 keeps the two tiles and play passes to player 2. If the expression and model are not equivalent, the tiles are replaced, and play passes to player 2. The game ends when all the tiles have been removed from the board. The player with more tiles wins.

	A	B	C	D
4	$3(x+3)$	[model: 2 x-tiles, 2 negative units]	[model: 6 x-tiles and 3 x-tiles]	$x-3$
3	$3(3x)$	[model: 1 x-tile, 5 positive units]	$3x+3$	[model: 4 x-tiles, 6 positive units]
2	[model: 1 x-tile, 3 negative units]	$4x+8$	[model: 3 x-tiles, 6 positive units]	$2x-2$
1	[model: 3 x-tiles, 2 positive units]	$2(2x-4)$	$x+5$	[model: 4 x-tiles, 8 negative units]

Key: ☐ represents positive values   ■ represents negative values

Unit 24 Homework

Standards 6.7(B) – Supporting, 6.7(C) – Supporting

1. Bennett and Larissa each rewrite the following expression:

    $a + a + 2a + 7 + 4 + 3a$

    Bennett writes $4a + 11 + 3a$, and Larissa writes $7a + 11$. Which student wrote an equivalent expression? Explain.

    _____

    _____

2. Write two different expressions to represent the model shown below.

    [Model: four x bars and eight unit squares labeled 1]

    _____

    _____

3. The table shows the prices of some items at the movie theater.

Popcorn	$3
Soda	$4
Candy	$2

    Write two different equivalent expressions to find the cost of purchasing 2 popcorns, 2 sodas, and 1 candy.

    _____

    _____

4. Classify each of the following as an expression or an equation by circling the correct response.

    a. A weekly allowance of $12.50

        Expression/Equation

    b. $23x + 15(10) = 610$

        Expression/Equation

    c. The product of 5 and a number is 10.

        Expression/Equation

    d. The area of a rectangle with a length of 8 inches and a width of w inches is 72 square inches.

        Expression/Equation

5. Which of the following expressions are equivalent? Justify your answer using a model.

    a. $6(10) + 2$

    b. $2(6 \cdot 5 + 1)$

    c. $3(20 + 1)$

Connections

1. Write a paragraph comparing the similarities and differences between expressions and equations in math with the similarities and differences between phrases and sentences in English language arts. Share your work with your math and language arts teachers.

2. Use a grocery ad to create a poster demonstrating how to write equivalent expressions. Choose two or three ads and the number of each item you wish to purchase. Then write equivalent expressions showing how to find the total amount that would be spent. Glue the pictures on a sheet of paper and label each with the corresponding expression.

Standard 6.7(D) – Readiness

Unit 25 Introduction

1 Use the identity property of addition to create an expression equivalent to -8.

_____

Use the identity property of multiplication to create an expression equivalent to -8.

_____

Explain the difference between the identity property of addition and the identity property of multiplication.

_____

_____

2 Dominic uses 44.3 MB of memory downloading a game and 56.1 MB of memory downloading a fitness app to his smart phone. Write an expression that can be used to find the total amount of memory Dominic uses on the two downloads.

_____

Write an equivalent expression using the commutative property of addition.

_____

Write a sentence that explains the commutative property of addition.

_____

_____

3 The Albrecht family purchased snacks at the movie theater. Each of the 5 family members bought a small bag of popcorn for $4.50 and a bottle of water for y dollars. Write two equivalent expressions using the distributive property that can be used to find the total amount of money the Albrecht family spent on snacks.

_____

_____

Write a sentence that explains the distributive property.

_____

_____

4 Look at the model below.

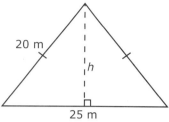

Generate two equivalent expressions using the associative property that can be used to find the area of the triangle.

_____

Write a sentence that explains the associative property of multiplication.

_____

_____

5 Use the inverse property of addition to complete the equation below.

$-\frac{1}{4} +$ _____ = _____

When using the inverse property of addition, the expression will always be equivalent to _____.

Use the inverse property of multiplication to complete the equation below.

$-5 \times$ _____ = _____

When using the inverse property of multiplication, the expression will always be equivalent to _____.

6 Complete the table to show if each pair of expressions is equivalent. If yes, identify which property is applied.

Expressions	Yes/No	Property
$(3 + b) \cdot 5$ $15 + 5b$		
$8 \cdot (-5 + 6)$ $(8 \cdot 5) + 6$		
$(17 + 3c)12$ $12(17 + 3c)$		

Unit 25 Guided Practice

Standard 6.7(D) – Readiness

1 Jake records the expression below.

$$b \cdot 6$$

Which of the following uses the commutative property to represent an equivalent expression?

Ⓐ $b \cdot b \cdot b \cdot b \cdot b \cdot b$

Ⓑ $1 \cdot b \cdot 6$

Ⓒ $b + b + b + b + b + b$

Ⓓ $6b$

2 Mariah uses the distributive property to write an expression equivalent to $18 + 36$. Which of the following best represents Mariah's expression?

Ⓕ $18(1 + 4)$

Ⓖ $9(2 + 4)$

Ⓗ $18(2 + 2)$

Ⓙ $6(2 + 6)$

3 Morgan earns $15 per hour tutoring. Last week she tutored for 6 hours. Morgan uses the expression $15 \cdot 6$ to determine the amount of money she earned tutoring last week. Which of the following uses the identity property of multiplication to show an expression that is equivalent to the expression Morgan wrote?

Ⓐ $(15 \cdot 6) \cdot 1$

Ⓑ $6 \cdot 15$

Ⓒ $(15 \cdot 6) \cdot 0$

Ⓓ $(15 \cdot 6) \cdot \frac{1}{90}$

4 Cecilia and Olivia are playing a game in which they match equivalent expressions. Cecilia matches a pair of equivalent expressions using the associative property of addition. Which best represents the cards Cecilia matches?

Ⓕ | $10 + (7n + 3)$ | $(7n + 3) + 10$ |

Ⓖ | $(7n + 3) + 10$ | $7n + (3 + 10)$ |

Ⓗ | $10 + (7n + 3)$ | $10(7n) + 10(3)$ |

Ⓙ | $7n + (3 + 10)$ | $7n + 13$ |

5 Which property best justifies why the two expressions below are equivalent?

$$5(w + 12) = 5w + 60$$

Ⓐ Associative property

Ⓑ Commutative property

Ⓒ Inverse property

Ⓓ Distributive property

6 Lauren generated equivalent expressions using the inverse property of multiplication. Which expression does NOT correctly apply this property?

Ⓕ $\frac{3}{4} \cdot \frac{4}{3}$

Ⓖ $-22 \cdot -\frac{1}{22}$

Ⓗ $\frac{2}{5} \cdot 1$

Ⓙ $\frac{1}{2} \cdot 2$

Standard 6.7(D) – Readiness

Unit 25 Independent Practice

1 Which number sentence correctly uses an identity property to show equivalent expressions?

Ⓐ $6.25(0) = 0$

Ⓑ $5\frac{3}{4} + 1 = 6\frac{3}{4}$

Ⓒ $-13 + 13 = 0$

Ⓓ $\frac{10}{2}(1) = 5$

2 A pasture with horses and goats has a width of $\frac{1}{4}$ mile. The length of the pasture is represented by $x + 8$. Which pair of expressions represents the area of the pasture and uses the distributive property to show equivalence?

Ⓕ $\frac{1}{4}(x + 8)$ and $(x + 8) \cdot \frac{1}{4}$

Ⓖ $\frac{1}{4}(x + 8)$ and $\frac{1}{4}x + 8$

Ⓗ $\frac{1}{4}(x + 8)$ and $\frac{1}{4}x + \frac{1}{4}(8)$

Ⓙ $\frac{1}{4}(x + 8)$ and $\frac{1}{4}(x + 8) \cdot 1$

3 Look at the expressions below.

Expression 1: $x - x$

Expression 2: $-2 + \frac{4}{2}$

Expression 3: $\frac{1}{4} + 0.25$

Expression 4: $-\frac{1}{5} - \left(-\frac{1}{5}\right)$

Zion made a mistake when using the inverse property of addition to generate equivalent expressions. Which expression shows the mistake Zion made?

Ⓐ Expression 1

Ⓑ Expression 2

Ⓒ Expression 3

Ⓓ Expression 4

4 Mrs. Hutchins displayed the following two expressions.

$(2t - 6) + (5 \cdot 4t)$ and $(5 \cdot 4t) + (2t - 6)$

Which statement is true about the expressions Mrs. Hutchins displayed?

Ⓕ The expressions are equivalent because of the associative property.

Ⓖ The expressions are equivalent because of the commutative property.

Ⓗ The expressions are equivalent because of the distributive property.

Ⓙ The expressions are not equivalent because they do not represent the same value.

5 Which property best explains the reason the two expressions below are equivalent?

$3(x + 8) = 3x + 3(8)$

Ⓐ Distributive property

Ⓑ Commutative property

Ⓒ Associative property

Ⓓ Not here

6 Nambi records the expression below.

$3d \cdot (6 \cdot -7)$

Which of the following uses the associative property to represent an equivalent expression?

Ⓕ $(6 \cdot -7) \cdot 3d$

Ⓖ $3d(6) + 3d(-7)$

Ⓗ $(3d \cdot 6) \cdot -7$

Ⓙ $3d \cdot (6 \cdot -7) \cdot 1$

Unit 25 Assessment

Standard 6.7(D) – Readiness

1 Mrs. Erskine spent $40.50 for a pair of shoes, $56.95 for jeans, and $25.75 for a shirt. Which of the following shows two equivalent expressions, using the associative property of addition, that can be used to find the amount Mrs. Erskine spent?

Ⓐ (40.50 + 56.95) + 25.75; 25.75 + (40.50 + 56.95)

Ⓑ 40.50 + (56.95 + 25.75); (40.50 + 56.95) + 25.75

Ⓒ 40.50 + (56.95 + 25.75); (40.50 • 56.95) + (40.50 • 25.75)

Ⓓ (40.50 + 56.95) + 25.75; (40.50 + 56.95 + 25.75) • 1

2 Kara sorted equivalent expressions into groups based on certain properties. Kara grouped the expressions shown below.

$$-x + x$$

$$1\tfrac{1}{2} + \left(-\tfrac{3}{2}\right)$$

$$0.3 + \left(-\tfrac{3}{10}\right)$$

Which statement is true about the expressions Kara grouped together?

Ⓕ The expression $-x + x$ does not belong because the other two expressions show the identity property of addition.

Ⓖ Kara correctly grouped the expressions showing the identity property.

Ⓗ The expression $1\tfrac{1}{2} + \left(-\tfrac{3}{2}\right)$ does not belong because the other two expressions show the inverse property of addition.

Ⓙ Kara correctly grouped the expressions showing the inverse property of addition.

3 Lo uses mental math to compute 7 × 92. She thinks (7 × 90) + (7 × 2). Which property does Lo use to mentally compute 7 × 92?

Ⓐ Commutative property

Ⓑ Distributive property

Ⓒ Inverse property of multiplication

Ⓓ Associative property

4 Mila saves $5 every week from her allowance. Which of the following shows two equivalent expressions, using the commutative property, that can be used to find the amount of money Mila saves in a year?

Ⓕ $5x = x + x + x + x + x$

Ⓖ 5 • 365 = (5 • 365) • 1

Ⓗ 5(52) = 52 • 5

Ⓙ 12(5) = 5(12)

5 Todd wrote equivalent expressions to demonstrate each of the properties shown below. Which property is NOT correctly demonstrated in the expressions Todd wrote?

Associative Property
$(7 - x) - 5 = 7 - (x - 5)$

Commutative Property
$(5x + 2) + x = x + (5x + 2)$

Distributive Property
$\tfrac{1}{5}(x + 20) = \tfrac{1}{5}(x) + \tfrac{1}{5}(20)$

Identity Property
$\tfrac{x}{10}(1) = \tfrac{x}{10}$

Ⓐ Associative property

Ⓑ Commutative property

Ⓒ Distributive property

Ⓓ Identity property

Standard 6.7(D) – Readiness   **Unit 25** Critical Thinking

1 Marco is adding two fractions that have different denominators. How can he use the identity property of multiplication to complete the computation? Include an example problem and solution in your explanation.

_____

_____

_____

Give another example of a math concept, other than adding fractions, where the identity property of multiplication is used.

2 Consider the following equation:

$$a \cdot \dfrac{1}{\frac{1}{2}} = 1$$

What property is demonstrated by the equation? _____

What value of *a* makes this equation true? Justify your response.

_____

_____

_____

Unit 25 Journal/Vocabulary

Name _____

Standard 6.7(D) – Readiness

Journal

Explain why the commutative and associative properties do not apply to subtraction or division.

_____
_____
_____

Vocabulary Activity

For each term, write a definition in your own words. Then give an example of two expressions that are equivalent and demonstrate the property.

Commutative property of addition or multiplication: _____
_____

Example: _____ = _____

Associative property of addition or multiplication: _____
_____

Example: _____ = _____

Distributive property: _____
_____

Example: _____ = _____

Additive identity property: _____
_____

Example: _____ = _____

Multiplicative identity property: _____
_____

Example: _____ = _____

Additive inverse property: _____
_____

Example: _____ = _____

Multiplicative inverse property: _____
_____

Example: _____ = _____

Property Roll

Play *Property Roll* with a partner. Each pair of players needs three number cubes, a game board, and a paper clip to use with the spinner. Each player needs a pencil. Player 1 spins the spinner, and then rolls the number of cubes indicated on the section spun. Player 1 writes two equivalent expressions using the property on the spinner and the numbers rolled on the cubes. If the two expressions are written correctly, player 1 earns points by adding the values on the number cubes rolled. Play then passes to player 2. If a player lands on a space already completed in his/her table, he/she loses a turn and play passes to the other player. The game ends when both players' tables are complete. The player with more points wins.

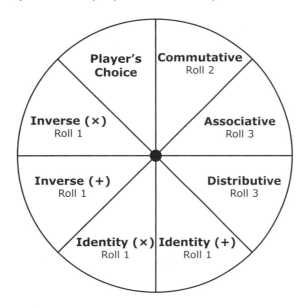

PLAYER ONE			PLAYER TWO		
Property	Expressions	Points Earned	Property	Expressions	Points Earned
Associative			Associative		
Commutative			Commutative		
Distributive			Distributive		
Identity (+)			Identity (+)		
Identity (×)			Identity (×)		
Inverse (+)			Inverse (+)		
Inverse (×)			Inverse (×)		

Unit 25 Homework

Standard 6.7(D) – Readiness

1 Shawna collected cans for 3 consecutive days. She collected 22 cans on Friday, 28 cans on Saturday, and 16 cans on Sunday. Using the associative property of addition, write two equivalent expressions that could be used to find the total number of cans Shawna collected in 3 days.

_____

_____

2 For each set of equivalent expressions, state the specific property that justifies each equivalence.

a. $4x - 8 = 4(x - 2)$

_____

b. $1 \cdot 7 = 7$

_____

c. $\left(\frac{1}{2}b\right)h = \frac{1}{2}(bh)$

_____

d. $3(4 + w) = 3(w + 4)$

_____

e. $\left(\frac{2}{3}\right)\left(\frac{3}{2}\right) = 1$

_____

f. $-2x + 0 = -2x$

_____

g. $\left(\frac{1}{12}\right)\left(\frac{4}{x}\right) = \left(\frac{4}{x}\right)\left(\frac{1}{12}\right)$

_____

h. $17 + (-17) = 0$

_____

Use the model to answer questions 3–5.

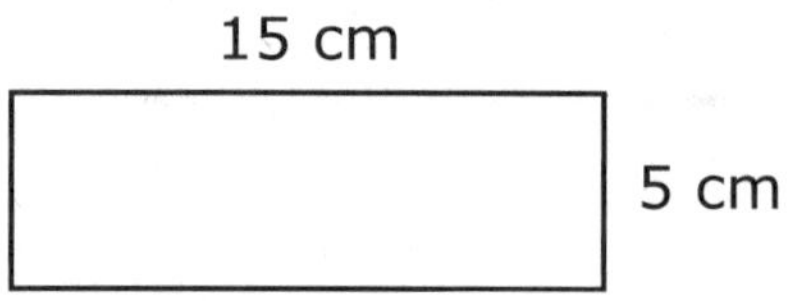

15 cm

5 cm

3 Using the commutative property of multiplication, write two equivalent expressions that could be used to find the area of the figure above.

_____

_____

4 Using the commutative property of addition, write two equivalent expressions that could be used to find the perimeter of the figure above.

_____

_____

5 Using the distributive property, write two equivalent expressions that could be used to find the perimeter of the figure above.

_____

_____

6 Apply the identity property of multiplication to the following problem to find common denominators. Then find the sum of the two fractions.

$$\frac{2}{3}$$
$$+ \frac{1}{2}$$
$$\overline{}$$

Connections

Create a set of cards to use as a game to help learn the operation properties. Using index cards, write one property on each card. On a separate card, write the definition of the property and an example. Play *Memory*, *Go Fish*, or another card game with a friend or family member to help you learn the properties.

Name _____

Standard 6.8(A) — Supporting

Unit 26 Introduction

1 Use a ruler to measure the lengths of the three line segments to the nearest $\frac{1}{2}$ inch.

―――――――――――――――――

――――――

――――――――――――

Can the lines be arranged to form a triangle?
_____

Justify your answer.

_____

_____

_____

2 Zhi needs to draw an example of a scalene triangle. Look at each set of side lengths or angle measures. For each set, determine if a scalene triangle can be formed. Explain your reasoning.

25°, 55°, 100°

_____

_____

7 cm, 10 cm, 3 cm

_____

_____

110°, 35°, 35°

_____

_____

$5\frac{1}{2}$ cm, $12\frac{3}{4}$ cm, 18 cm

_____

_____

64°, 48°, 58°

_____

_____

3 Use the clues below to answer the questions about △JKL.

- The measure of ∠J (m∠J) is equal to the product of 9 and 4.
- The measure of ∠K (m∠K) is $\frac{3}{2}$ the measure of ∠J.

Sketch and label △JKL.

What is the measure of ∠J? _____

What is the measure of ∠K? _____

Using this information, how can you determine the measure of ∠L?

_____

_____

What is the measure of ∠L? _____

Which side of the triangle is the longest? How do you know?

_____

_____

Which side of the triangle is the shortest? How do you know?

_____

_____

Unit 26 Guided Practice

Standard 6.8(A) – Supporting

1. Gracelyn has multiple sticks of different lengths to build a triangle for a class project. Which 3 lengths could Gracelyn choose to build the triangle?

   Ⓐ 10 cm, 22 cm, 4 cm

   Ⓑ 12 cm, 5 cm, 17 cm

   Ⓒ 13 cm, 6 cm, 6 cm

   Ⓓ 11 cm, 25 cm, 15 cm

2. Study the triangle below.

   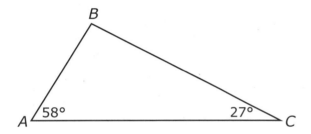

   Which statement about △ABC is NOT true?

   Ⓕ $\overline{BC}$ is the longest side because it is opposite the largest angle.

   Ⓖ △ABC is an obtuse scalene triangle.

   Ⓗ $\overline{AB}$ is the shortest side because it is opposite the smallest angle.

   Ⓙ ∠B measures 95° because the sum of the angles in a triangle is 180°.

3. Lamont builds a triangular shaped fence to hide his pool pump. He already has two sections of fencing measuring $8\frac{1}{2}$ feet and $5\frac{1}{2}$ feet. Which fencing length would Lamont NOT be able to use as the third side of his fence?

   Ⓐ 4 ft

   Ⓑ 8 ft

   Ⓒ 12 ft

   Ⓓ 16 ft

4. According to the diagram, how would you find the measure of ∠B?

   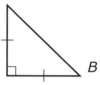

   Ⓕ Subtract 90° from 180°

   Ⓖ Subtract 180° from the sum of 45° and 90°

   Ⓗ Subtract 90° from 180°, and then divide the difference by 2

   Ⓙ Subtract 180° from the sum of 45° and 45°

5. Triangle JKL is shown below.

   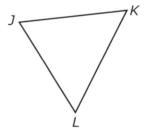

   If KL = 12 centimeters, JK = 10 centimeters, and JL = 11 centimeters, which list shows the angles in order from least to greatest, according to their measures?

   Ⓐ ∠J, ∠K, ∠L

   Ⓑ ∠L, ∠K, ∠J

   Ⓒ ∠L, ∠J, ∠K

   Ⓓ ∠K, ∠L, ∠J

Standard 6.8(A) – Supporting

Unit 26 Independent Practice

1. Look at the figure below.

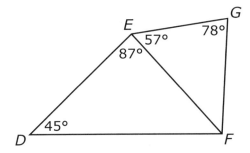

Find the measure of ∠DFG based on the given information in the diagram.

Record your answer and fill in the bubbles on the grid below. Be sure to use the correct place value.

2. Josefina decorates the corner of a page in her scrapbook with different lengths of ribbon. One blue ribbon measures 23 centimeters, and one red ribbon measures 14 centimeters. If Josefina makes a triangle with the ribbon lengths, which of the following represents a possible length of the third ribbon?

   I. 8 cm
   II. 12 cm
   III. 36 cm
   IV. 40 cm

   Ⓕ I only
   Ⓖ II and IV only
   Ⓗ II and III only
   Ⓙ III and IV only

3. Triangle XYZ is shown below.

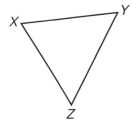

If XY = 7.8 inches, YZ = 8 inches, and XZ = 7.75 inches, which statement is true?

Ⓐ ∠Z is the smallest angle measure.
Ⓑ ∠Y is the smallest angle measure.
Ⓒ ∠Z is the largest angle measure.
Ⓓ ∠X is the smallest angle measure.

4. Gwen forms a right triangle for a wire sculpture. The measure of the smallest angle is $\frac{1}{3}$ the measure of the largest angle. The third angle has a measure that is twice as large as the measure of the smallest angle. What are the angle measures of Gwen's triangle?

Ⓕ 30°, 60°, 90°
Ⓖ 45°, 90°, 135°
Ⓗ 20°, 40°, 60°
Ⓙ 45°, 45°, 90°

5. Eva is running for class president. She creates a sign in the shape of an isosceles triangle to hang in the cafeteria. If the base of the triangle is 36 inches, which of the following is true about the other sides?

Ⓐ One side measures 24 inches and the other 20 inches.
Ⓑ The other sides must each measure 36 inches.
Ⓒ The other sides must be equivalent and each measure greater than 18 inches.
Ⓓ The other sides must be equivalent and each measure less than 15 inches.

Unit 26 Assessment

Standard 6.8(A) – Supporting

1. Nat sews a triangle on an apron. The three sides of the triangle measure 3 inches, 4 inches, and 5 inches. Which best represents the triangle Nat sews on her apron?

   Ⓐ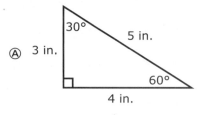

   Ⓑ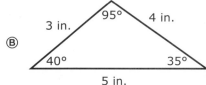

   Ⓒ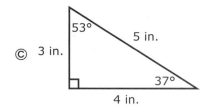

   Ⓓ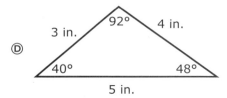

2. In triangle *XYZ*, the measure of ∠Y is 64°. The measure of ∠Z is $\frac{3}{4}$ the measure of ∠Y. What is the measure of ∠X?

   Ⓕ 112°

   Ⓖ 116°

   Ⓗ 68°

   Ⓙ 48°

3. Nathan cut an isosceles triangle from felt to make a spirit banner. Two sides of his banner had the following measures: 15 inches and 7 inches. Which could be the measure of the third side of Nathan's banner?

   Ⓐ 20 inches

   Ⓑ 15 inches

   Ⓒ 10 inches

   Ⓓ 7 inches

4. The drawings below show different triangles and information about some of their angle measures.

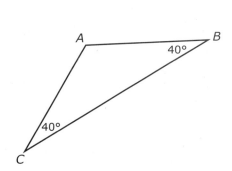 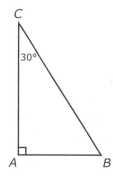 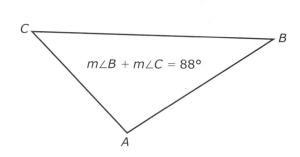

Which of the following statements is true?

Ⓕ ∠C represents the smallest angle in all three triangles.

Ⓖ All of the triangles are scalene.

Ⓗ m∠A > 90° in all three triangles.

Ⓙ $\overline{BC}$ is the longest side in all three triangles.

Name _____

Standard 6.8(A) – Supporting

Unit 26 Critical Thinking

1. Record the dimensions shown in the box in the appropriate location on the triangle.

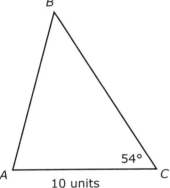

| 10 units | 54° | 18 units | 72° |

Note: Triangle not drawn to scale.

Explain how you solved the problem and justify your answer.

_____
_____
_____
_____

2. Shelby draws and labels the following triangle.

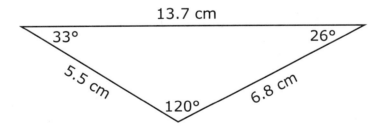

Explain the mistakes Shelby made in her drawing. Then, explain possible changes to correct the mistakes.

_____
_____
_____
_____

Unit 26 Journal/Vocabulary Activity

Name _____

Standard 6.8(A) – Supporting

Journal

Explain the following statement: *The two acute angles of a right triangle are complementary.*

_____
_____
_____
_____

Vocabulary Activity

In this unit, you have learned about three important properties of triangles. These properties have special names called *theorems*. Match the name of the theorem with its definition and illustration. Circle each set of matches with a different color.

THEOREM NAME

| Triangle Sum Theorem | Triangle Inequality Theorem | Opposite Side-Angle Theorem |

DEFINITION

| If one side of a triangle is longer than another side, then the angle opposite the longer side is larger than the angle opposite the shorter side. | The sum of the measures of the angles in a triangle is 180°. | The sum of the lengths of any two sides of a triangle is greater than the length of the third side. |

ILLUSTRATION

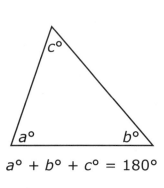

$a° + b° + c° = 180°$

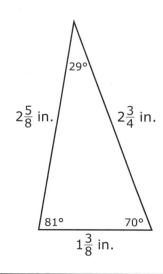

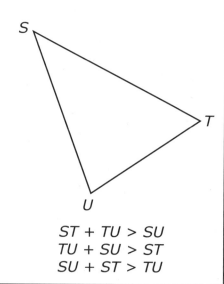

ST + TU > SU
TU + SU > ST
SU + ST > TU

212   ILLEGAL TO COPY   motivation**math** LEVEL 6   ©2014 mentoringminds.com

Name _____

Standard 6.8(A) – Supporting

Unit 26 Motivation Station

Does it Fit?

Play *Does it Fit?* with a partner. Each pair of players needs three small tokens and a game board. Each player needs a pencil. Player 1 begins by randomly tossing the three tokens onto the game board. After recording the three lengths the tokens land on in the chart, the player must determine whether a triangle can be formed. If the answer is *yes*, the player writes *yes* in the box and earns 5 points. If the answer is *no*, the player writes *no* in the box and earns -2 points. The game continues until each player has had ten turns. The player with more points wins.

3 units	$\frac{3}{4}$ unit	11 units	8.5 units	12 units
5.7 units	$\frac{1}{2}$ unit	13 units	9 units	7 units
6 units	2 units	20 units	4 units	$\frac{2}{3}$ unit
5 units	1 unit	2.9 units	10 units	8 units

Player 1

Player 2

Length 1	Length 2	Length 3	Yes/No	Points	Length 1	Length 2	Length 3	Yes/No	Points
TOTAL POINTS					TOTAL POINTS				

Unit 26 Homework

Standard 6.8(A) – Supporting

1 Five possible side lengths for a triangle are shown below.

———— 1.4 in.

———————— 2.2 in.

——————— 0.9 in.

———————— 2 in.

——————————— 2.7 in.

Select 3 side lengths to form a triangle. Justify your selections.

_____

_____

2 Triangle STU is shown below.

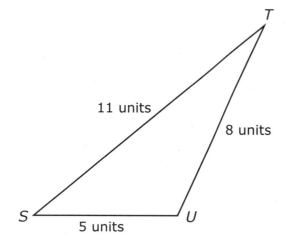

List the angles in order from least to greatest, according to their measure.

_____

3 A triangle has 2 congruent angles, and the third angle is acute. List two possible sets of angle measures for the triangle.

_____

_____

4 Mrs. Colfax plans to build a triangular garden in her backyard. She has 25 feet of chicken wire to put around the garden. She wants one side of the garden to measure 14 feet and the second side to measure 4 feet. The remainder of the wire will be used on the third side. Will the dimensions Mrs. Colfax has planned form a triangle? Explain.

_____

_____

_____

5 Isosceles triangle ABC is shown.

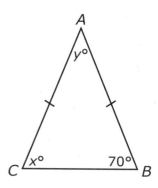

What is the value of x? _____

What is the value of y? _____

Connections

1. Explore occupations that use triangles, such as construction or surveying. Make a list of five different occupations and explain how triangles are used. Share your findings with the class.

2. Create a card game to practice finding missing angles in a triangle. Draw different triangles with two angles labeled on one side of the card, and the third angle measure written on the back. Ask someone to show you the triangle side of the card and you answer with the measure of the third angle.

Name _____

Standard 6.8(B) – Supporting   Unit 27 Introduction

1. Simon has a rectangular piece of fabric. He cuts it to create 2 congruent right triangular flags.

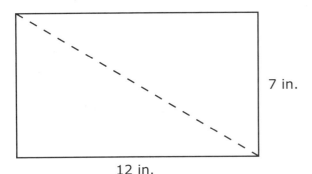

Explain how to find the area of one flag.

_____

_____

Create a formula to find the area of the triangular flag.

_____

How many square inches of fabric will Simon use to create one flag?

_____

2. Kwan needs to find the area of the parallelogram shown below.

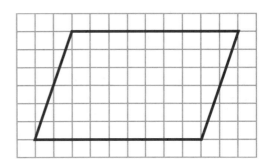

He decides to remove the triangular piece from one end and place it on the other end to create a rectangle. Use the picture above to sketch what Kwan did.

How can you determine the area of the rearranged parallelogram?

_____

_____

Write a formula to find the area of the parallelogram.

_____

Find the area of the parallelogram, in square units. _____

3. Jessie cuts an isosceles trapezoid down the middle and rearranges the pieces to create a rectangle, as shown in the picture below.

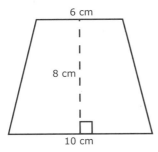 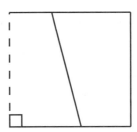

Write an expression to find the area of the rectangle. _____

The formula for finding the area of a trapezoid is $A = \frac{1}{2}(b_1 + b_2)h$. Using the values from the trapezoid above, write an expression to find the area. _____

Explain how the expressions for the rectangle and the trapezoid are alike and different.

_____

_____

©2014 mentoringminds.com   motivationmath™ LEVEL 6   ILLEGAL TO COPY   215

Unit 27 Guided Practice

Standard 6.8(B) – Supporting

1 Makayla wants to find the area of the parallelogram shown below.

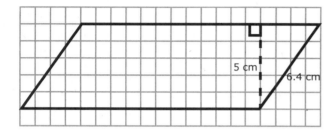

Which of the following does NOT describe how Makayla can use the model to find the area of the parallelogram?

Ⓐ Estimate the number of squares contained in the figure.

Ⓑ Remove a triangular piece from one end of the figure and place it on the opposite end.

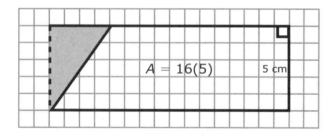

Ⓒ Multiply the dimensions of the figure, $A = 16(6.4)$.

Ⓓ Draw a diagonal on the figure, and find the area of one of the resulting triangles. Multiply the area by 2 to find the area of the parallelogram.

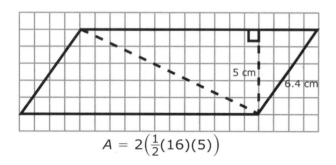

$A = 2\left(\frac{1}{2}(16)(5)\right)$

2 Mrs. Lang asked the class to find the area of an isosceles trapezoid. Sonya decomposed the trapezoid and rearranged the parts to form a rectangle of equal area, as shown.

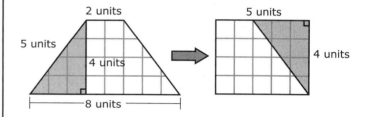

Which pair of expressions represents the area of the trapezoid?

Ⓕ $\frac{1}{2}(5 \times 4)$ and $(8 - 3)(4)$

Ⓖ $\frac{1}{2}(2 + 8)(4)$ and $(2 + 3)(4)$

Ⓗ 8×5 and $2(5)(4)$

Ⓙ $2(8 + 4)$ and $(8 - 2)(4)$

3 Lin created a design in his scrapbook by cutting square sheets of paper along the diagonal, as shown below.

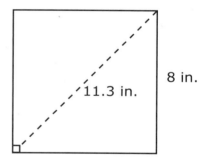

Which equation can be used to determine the area of one triangle?

Ⓐ $A = (8)(8)$

Ⓑ $A = (8)(11.3)$

Ⓒ $A = \dfrac{(8)(8)}{2}$

Ⓓ $A = \dfrac{(8)(11.3)}{2}$

Standard 6.8(B) – Supporting

Unit 27 Independent Practice

1. Which of the following could NOT be used to determine the area of the parallelogram by rearranging parts to create a rectangle?

Ⓐ

Ⓑ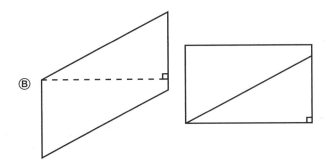

Ⓒ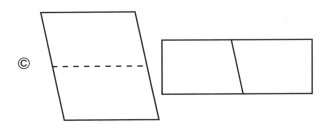

Ⓓ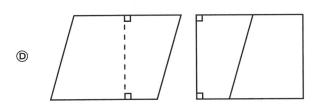

2. Raj is planting ivy as a ground cover. The portion of his yard to be planted in ivy is shaped like a trapezoid and is shown in the sketch below.

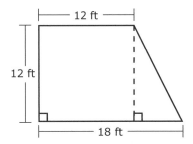

Raj does not remember how to find the area of a trapezoid. Which equation could Raj use to determine the area of his yard that will be planted in ivy?

Ⓕ $A = (12 \cdot 12) + \left(\dfrac{6 \cdot 12}{2}\right)$

Ⓖ $A = (12 \cdot 12) + \left(\dfrac{18 \cdot 12}{2}\right)$

Ⓗ $A = (12 \cdot 12) + \left(\dfrac{4 \cdot 12}{2}\right)$

Ⓙ $A = (12 \cdot 12) + (6 \cdot 12)$

3. Suzi plans quilt designs using isosceles triangles cut from construction paper. Suzi cuts her isosceles triangle along the dotted line to form two halves and rearranges the pieces to form a rectangle. Suzi's work is shown below.

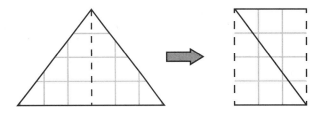

Which equation can be used to find the area of Suzi's triangle?

Ⓐ $A = 6 \times 5$ Ⓒ $A = \frac{1}{2}(6)(4)$

Ⓑ $A = 6 \times 4$ Ⓓ $A = \frac{1}{2}(6)(5)$

Unit 27 Assessment

Standard 6.8(B) – Supporting

1. Lonell uses tiles in the shape of trapezoids to create a backsplash for the kitchen counter. He arranges two trapezoids to form a parallelogram with dimensions marked, as shown below.

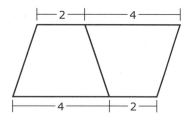

Lonell knows that the area of a parallelogram can be found using the equation $A = bh$. If the height of the parallelogram is 3 units, which equation can be used to find the area of each trapezoid?

Ⓐ $A = \dfrac{(4 + 2)(3)}{2}$

Ⓑ $A = 4 \times 2 \times 3$

Ⓒ $A = \dfrac{(4 \times 2 \times 3)}{2}$

Ⓓ $A = 3(4 + 2)$

2. Joey wants to find the area of a parallelogram. He decomposes the parallelogram and rearranges the parts to form a rectangle of equal area, as shown below.

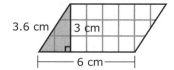 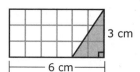

Which equation can be used to find the area of the parallelogram?

Ⓕ $6(3.6) = 21.6$ cm²

Ⓖ $2(3.6) + 2(6) = 19.2$ cm²

Ⓗ $6(3) = 18$ cm²

Ⓙ $6(3 + 3.6) = 39.6$ cm²

Use the information below to answer questions 3 and 4.

Paco creates a large triangle using an isosceles trapezoid and a small isosceles triangle.

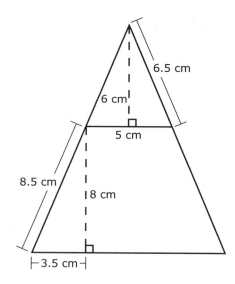

3. Which equation can Paco use to determine the area of the trapezoid?

Ⓐ $A = (3.5 + 5 + 3.5)(8)$

Ⓑ $A = (5 + 3.5)(8)$

Ⓒ $A = \frac{1}{2}(5 + 3.5)(8)$

Ⓓ $A = \frac{1}{2}(3.5 + 5 + 3.5)(8.5)$

4. Which best shows how to find the area of the triangle Paco created using the two figures?

Ⓕ $(3.5 + 5 + 3.5)(8) + \frac{1}{2}(5)(6)$

Ⓖ $(5)(6) + (3.5 + 5 + 3.5)(8)$

Ⓗ $(5 + 3.5)(8) + \frac{1}{2}(5)(6)$

Ⓙ $(5)(6) + (5 + 3.5)(8)$

Name _____

Standard 6.8(B) – Supporting

Unit 27 Critical Thinking

1 Study the following diagrams.

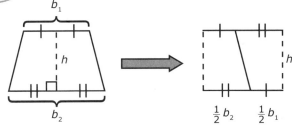

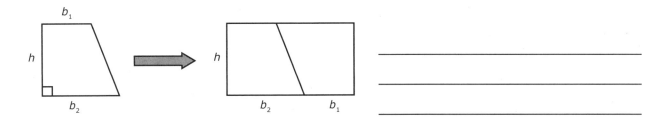

_____
_____
_____

_____
_____
_____

Beside each diagram, show how the formula, $A = \frac{1}{2}(b_1 + b_2)h$ or $A = \frac{(b_1 + b_2)h}{2}$ is derived, using the area formula for a rectangle and math properties.

2 June writes the following statement about triangles:

Every triangle can be duplicated and rearranged to create a rectangle.

Is June's statement always true, sometimes true, or never true? _____

Use pictures and words to justify your answer.

Unit 27 Journal/Vocabulary Activity

Name _____

Standard 6.8(B) – Supporting

Journal

Compare the formulas for the area of a rectangle and the area of a parallelogram. Use a diagram of each figure to demonstrate the similarities and differences.

Vocabulary Activity

Circle the *base(s)* of each of the following figures.

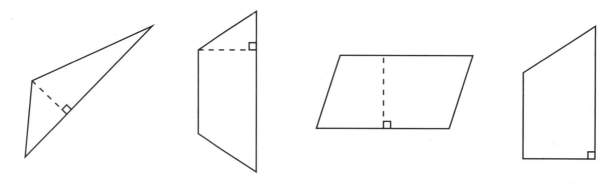

Circle the *height* of each of the following figures.

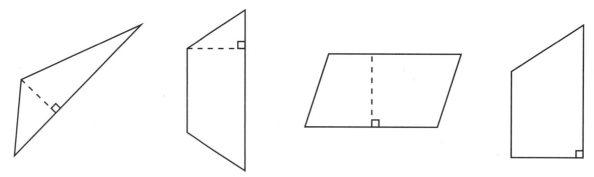

Circle the *altitude* of each of the following figures.

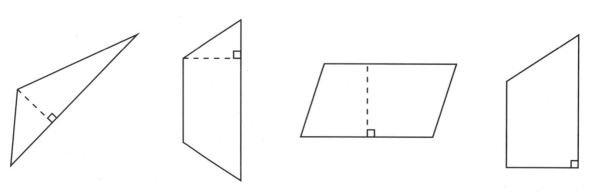

Name _____

Standard 6.8(B) – Supporting

Unit 27 Motivation Station

Analyze the Area

Play *Analyze the Area* with a partner. Each pair of players needs a game board and a paper clip to use with the spinner. Each player needs a game token, a pencil, and a sheet of paper. Player 1 spins the spinner and moves the game token the number of spaces indicated. Player 1 identifies the figure in the space and uses the dimensions to record the correct expression for finding the area of the figure. If recorded correctly, player 1 earns the points indicated in the space. If incorrect, player 1 scores no points. Play passes to player 2, who repeats the process. The game ends when both players reach the end of the game board path. The player with more points wins.

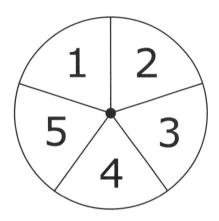

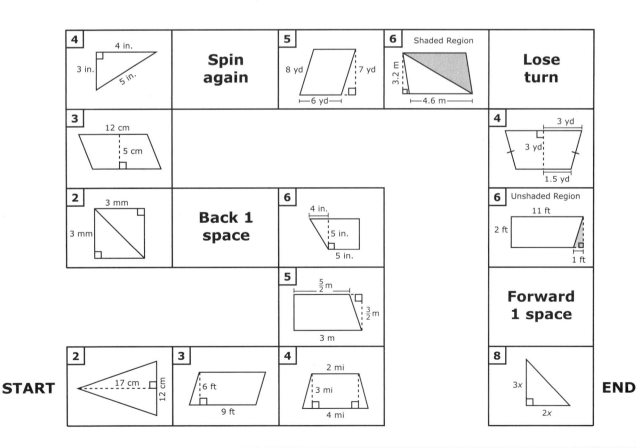

©2014 mentoringminds.com — motivationmath™ LEVEL 6 — ILLEGAL TO COPY — 221

Unit 27 Homework

Name _____

Standard 6.8(B) – Supporting

1. Mr. Hampton displays the following illustration.

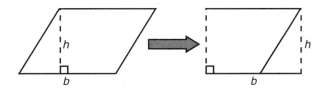

What is Mr. Hampton demonstrating in his drawing?

_____

_____

How is this information useful in finding the area of the parallelogram?

_____

_____

2. A parallelogram-shaped sheet of plywood is cut into 2 triangles, along a diagonal.

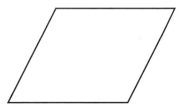

Use the diagram to sketch a line that can be used to form the triangles.

If the base of the parallelogram is 3 feet and the altitude of the parallelogram is 1.75 feet, write an equation that could be used to find the area of one triangle.

_____

Use the figure to answer questions 3–5.

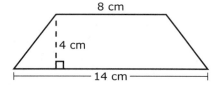

3. Use the space below to decompose and rearrange the isosceles trapezoid to form a rectangle with the same area. Label the dimensions of the rectangle.

4. Write an equation that can be used to find the area of the original figure.

_____

5. Use the space below to draw another way to decompose the trapezoid to find the area. Show how you would find the area using the new decomposition.

Connections

Select one shape: parallelogram, trapezoid, or triangle. Look through magazines or newspapers, or browse the Internet to print real-life examples of your shape. Cut out the examples and glue or tape them to a sheet of paper. Use a marker to draw lines on the shape to show how it can be decomposed, or duplicate the shape so it forms another figure. Below the examples, explain how the area of the original shape can be found.

Name _____

Standards 6.8(C) – Supporting, 6.8(D) – Readiness

Unit 28 Introduction

1. Foster had two slices of homemade pizza. One was cut like a rectangle and the other in the shape of a triangle, as shown below.

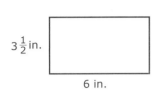
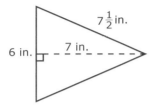

Write and solve an equation to determine the area of the rectangular slice of pizza, in square inches.

_____

Write and solve an equation to determine the area of the triangular slice of pizza, in square inches.

_____

How do the areas of the two slices of pizza compare?

_____

2. Mrs. Lynn purchases a new table for her classroom. The tabletop is shown in the model below.

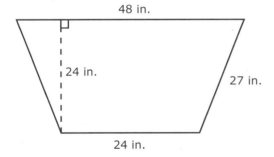

Write an equation that can be used to find the area of the tabletop, in square inches.

_____

What is the area of the tabletop, in square inches?

_____

3. Pedro constructs the figure below out of a cereal box.

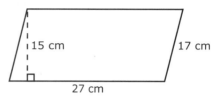

Write an equation that can be used to determine the area of the figure Pedro constructed, in square centimeters.

_____

Find the area of Pedro's figure, in square centimeters.

_____

4. Ellen's little brother, Jordan, plays with colored cubes that are stored in a rectangular box. The edge of each cube measures 3 inches. The dimensions of the storage box are shown below.

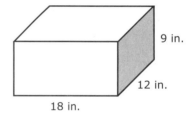

What is the volume of the storage box, in cubic inches?

_____

How many of Jordan's cubes will fit in the storage box?

_____

Explain how you found your answer.

_____

 **Unit 28** Guided Practice

Name _____

Standards 6.8(C) – Supporting, 6.8(D) – Readiness

1. Martin needs 108 square inches of metal to make a yield sign. If the height of the sign is 12 inches, how long is the top edge of the sign?

Ⓐ 9 in.

Ⓑ 18 in.

Ⓒ 54 in.

Ⓓ 96 in.

2. Lily has a sandbox that holds colored sand. The diagram below shows the dimensions of the sandbox.

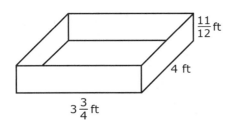

Lily's parents fill her sandbox to half its capacity. Which equation can be used to find the volume of the sand in Lily's sandbox?

Ⓕ $V = 4\left(\frac{15}{4}\right)\left(\frac{11}{12}\right)$

Ⓖ $V = \frac{1}{2}\left(4 + 3\frac{3}{4} + \frac{11}{12}\right)$

Ⓗ $V = \frac{1}{2}\left(4 \cdot 3\frac{3}{4} \cdot \frac{11}{12}\right)$

Ⓙ Not here

3. Use a ruler to measure the dimensions of the figure below to the nearest centimeter.

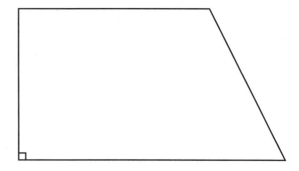

Which represents the number of square centimeters covered by the figure?

Ⓐ 48 cm²

Ⓑ 24 cm²

Ⓒ 54 cm²

Ⓓ 27 cm²

4. Levi draws the figures below.

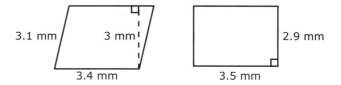

Which statement is true about the areas of the figures?

Ⓕ The area of the rectangle is 0.95 mm² larger than the area of the parallelogram.

Ⓖ The area of the parallelogram is 0.39 mm² larger than the area of the rectangle.

Ⓗ The area of the rectangle is 0.13 mm² larger than the area of the parallelogram.

Ⓙ The area of the parallelogram is 0.05 mm² larger than the area of the rectangle.

Name _____

Standards 6.8(C) – Supporting, 6.8(D) – Readiness

Unit 28 Independent Practice

1 Jimmy filled his fish aquarium with water one night.

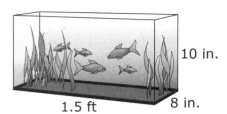

The next morning, Jimmy noticed the water level had dropped 1.5 inches. Which of the following equations could Jimmy use to determine the volume of water, in cubic inches, in his aquarium that morning?

Ⓐ $V = (18)(8)(10)$

Ⓑ $V = [(1.5)(8)(10)] - 1.5$

Ⓒ $V = (18)(8)(8.5)$

Ⓓ $V = (10)(8)$

2 Yazmine makes doll clothes for a sewing project. She uses the pattern below to make the front of a skirt for her sister's doll.

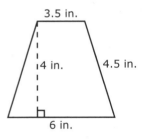

How many square inches of fabric will Yazmine use for the skirt pattern?

Ⓕ 21 in.²

Ⓖ 19 in.²

Ⓗ 25.5 in.²

Ⓙ 38 in.²

3 Gerald creates a kite to fly in the park. He decides to make the kite by attaching 2 congruent triangles at their bases. The pattern he uses for constructing one of the congruent triangles is shown below.

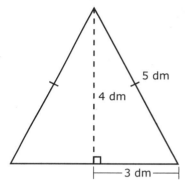

Which equation can Gerald use to determine the amount of material, in square decimeters, needed to construct the kite?

Ⓐ $A = \left[\frac{1}{2}(6)(4)\right] \cdot 2$

Ⓑ $A = \left[\frac{1}{2}(3)(4)\right] \cdot 2$

Ⓒ $A = [(6)(4)] \cdot 2$

Ⓓ $A = [(3)(5)] \cdot 2$

4 Lynette is covering shapes with wrapping paper to make a design for the school carnival.

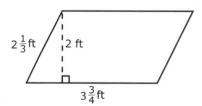

How much paper, in square feet, will Lynette need to cover the figure shown above?

Ⓕ $6\frac{3}{4}$ ft²

Ⓖ $7\frac{1}{2}$ ft²

Ⓗ $8\frac{3}{4}$ ft²

Ⓙ $7\frac{3}{4}$ ft²

Unit 28 Assessment

Standards 6.8(C) – Supporting, 6.8(D) – Readiness

1 Jasmine cut a parallelogram from a 4 × 6 inch index card, as shown below.

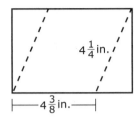

Which equation can Jasmine use to determine the area of the figure she cut?

Ⓐ $A = \left(4\frac{3}{8}\right)\left(4\frac{1}{4}\right)$

Ⓑ $A = \left(4\frac{3}{8}\right)(4)$

Ⓒ $A = \left(4\frac{3}{8}\right)(6)$

Ⓓ $A = \left(4\frac{1}{4}\right)(4)$

2 Rhett uses storage containers like the one shown below to organize the garage. If the volume of a storage container is 6,720 cubic inches, what is the height of the container?

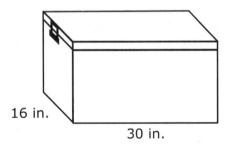

Ⓕ 12 in. Ⓗ 14 in.

Ⓖ 13 in. Ⓙ 15 in.

3 Mike wants to paint one wall of his room. The wall measures 12 feet by 9 feet. On the wall is one square window that measures $4\frac{1}{2}$ feet on each side. Which of the following best describes the number of square feet Mike will paint?

Ⓐ $87\frac{3}{4}$ ft²   Ⓒ 99 ft²

Ⓑ $103\frac{1}{2}$ ft²   Ⓓ Not here

4 Josie needs to find the area of the quadrilateral shown below.

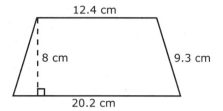

Which equation could Josie use to find A, the area of the quadrilateral?

Ⓕ $A = (20.2)(8)$

Ⓖ $A = \frac{1}{2}(12.4 + 20.2)(9.3)$

Ⓗ $A = \frac{1}{2}(20.2) + \frac{1}{2}(12.4)$

Ⓙ $A = \frac{1}{2}(20.2 + 12.4)(8)$

5 The Pet Palace is replacing its dog training arena with a new arena with dimensions that are twice as large. The old arena is shown below.

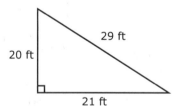

How much greater is the area of the new arena compared to the area of the old arena?

Ⓐ 630 ft²

Ⓑ 210 ft²

Ⓒ 1,260 ft²

Ⓓ 840 ft²

Name _____

Standards 6.8(C) – Supporting, 6.8(D) – Readiness

Unit 28 Critical Thinking

1. Chef Meredith has a window box in her kitchen that she uses to grow herbs for cooking. The box is pictured below.

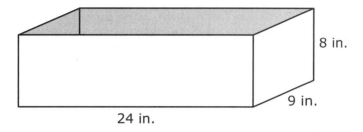

Potting soil is sold in bags with 2 cubic feet of soil in each bag. How many bags of soil will Chef Meredith use to completely fill her window box? Explain how you found your answer.

_____

_____

2. Sketch and label each of the following figures. Then write an equation that could be used to find the area of the figure.

- A trapezoid with an area of 100 square units

- A triangle with an area of 47.5 square units

- A parallelogram with an area of $33\frac{3}{4}$ square units

Unit 28 Journal/Vocabulary Activity

Standards 6.8(C) – Supporting, 6.8(D) – Readiness

Journal

Imagine you are given a figure composed of a rectangle with congruent triangles on each end. Draw a picture of how the figure might look and explain how to find the area of the entire figure.

Vocabulary Activity

Complete this activity with a partner. Each pair of students needs one 6-sided number cube. Each student needs a sheet of paper and a pencil. Take turns rolling the number cube and following the directions that match the number you rolled.

- ⚀ Sketch a rectangle and give an example of one located in the classroom.

- ⚁ Explain how to find the height or altitude of a parallelogram or trapezoid.

- ⚂ Describe how to find the area of any triangle.

- ⚃ Sketch a rectangular prism and label its length, width, and height.

- ⚄ Write the definition of a trapezoid.

- ⚅ Define volume and area. Give a real-world example of each.

Name _____

Standards 6.8(C) – Supporting, 6.8(D) – Readiness

Unit 28 Motivation Station

Achieve Area Authority

Play *Achieve Area Authority* with the class. The teacher begins the game by signaling start. Each player finds the area of the figures below and ranks them from greatest area to least area, with the rank of 1 noting the greatest area and 12 noting the least area. The first player to correctly rank the areas of the figures achieves area authority!

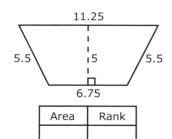

Area	Rank

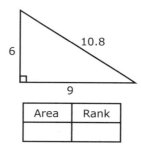

Area	Rank

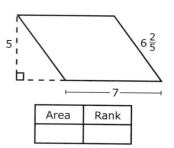

Area	Rank

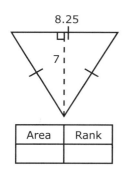

Area	Rank

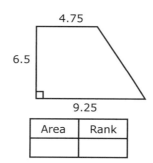

Area	Rank

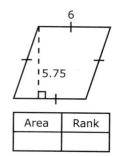

Area	Rank

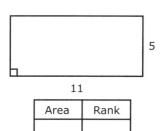

Area	Rank

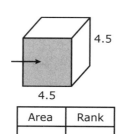

Area	Rank

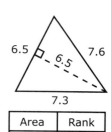

Area	Rank

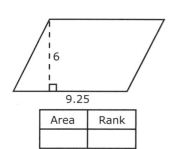

Area	Rank

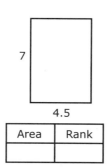

Area	Rank

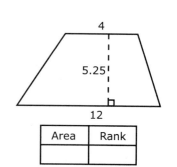

Area	Rank

©2014 mentoringminds.com   motivationmath™ LEVEL 6   ILLEGAL TO COPY

Unit 28 Homework

Standards 6.8(C) – Supporting, 6.8(D) – Readiness

1. A juice box measures 2.5 inches wide, 4 inches high, and 1.1 inches deep. Write an equation to find the amount of juice, in cubic inches, in one juice box.

   _____

   How much juice, in cubic inches, does a six-pack of juice boxes hold?

2. Shania creates an art piece like the one shown below, using red and black construction paper.

   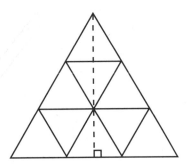

   The altitude of the art piece measures 18 inches, and the base of the piece is 15.6 inches. Five of the congruent triangles are made of red construction paper, and 4 are made of black construction paper. How many square inches of each color did Shania use to create her art piece? Show all work.

3. For each figure shown, write an equation to find the area of the figure. Then calculate the area.

   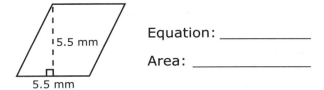

   Equation: _____

   Area: _____

   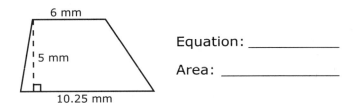

   Equation: _____

   Area: _____

   6 mm, 5 mm, 10.25 mm (trapezoid)

   Equation: _____

   Area: _____

4. The base of a rectangular prism has an area of 45 square centimeters. If the volume of the prism is 675 cubic centimeters, what is the height of the prism, in centimeters? Show all work.

Connections

Make plans to redecorate your bedroom. Measure the dimensions of your walls, floor, windows, and door(s). Calculate how many square feet of flooring you will need, how many square feet of wall space you will paint, without windows or doors, and how many square feet of doors you will paint, both sides. Research costs of paint and flooring and create a budget for your new bedroom. Share your budget with your parents or a friend.

Standards 6.9(A) – Supporting, 6.9(B) – Supporting

Unit 29 Introduction

For problems 1–5, write an equation or inequality to represent each situation within the set of positive rational numbers. Then graph the solution on the number line provided.

1. Vanessa takes $40 on her trip to the mall. Represent Vanessa's possible spending, s.

    _____

2. Sara is 11 years old today. Represent a, Sara's age in years.

    _____

    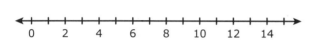

3. Kaitlyn served fewer than 15 cupcakes at her birthday party. Represent the number of cupcakes, c, Kaitlyn served at her party.

    _____

    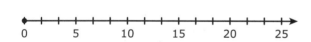

4. The post office receives a minimum of 8 pounds of mail per day. Represent the number of pounds, p, of mail the post office receives each day.

    _____

    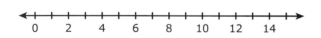

5. Lucas needs a larger hamster cage. The one he has now measures $\frac{3}{4}$ cubic foot. Represent the volume, v, of a new cage.

    _____

    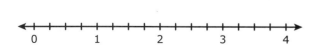

Use the information below to answer questions 6–8.

Bria takes piano lessons. She needs to practice at least 2 hours per week. So far this week she has practiced 95 minutes.

6. Will this situation be represented by an equation or an inequality?

    _____

    Explain. _____
    _____

7. Write an equation or inequality to determine the number of minutes, m, Bria still needs to practice this week.

    _____

8. Use the number line below to represent the solution to the equation or inequality written in question 7.

Use the information below to answer questions 9–11.

Max earns $8.50 per hour working at his father's shop. Last week he earned $136.

9. Will this situation be represented by an equation or an inequality?

    _____

    Explain. _____
    _____

10. Write an equation or inequality to determine the number of hours, h, Max worked last week.

    _____

11. Use the number line below to represent the number of hours Max worked last week.

    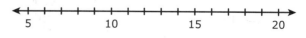

Unit 29 Guided Practice

Standards 6.9(A) – Supporting, 6.9(B) – Supporting

1 During Maleek's basketball game, he scored 37 points for his team. The rest of his teammates scored a total of s points. If the team won the game with 108 points, which equation could be used to determine the points earned by Maleek's teammates?

Ⓐ $\frac{108}{37} = s$

Ⓑ $37s = 108$

Ⓒ $s - 37 = 108$

Ⓓ $37 + s = 108$

4 Angles A and B are supplementary angles. If angle B measures 38°, which solution best represents the measure of angle A?

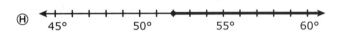

2 Nicole charges $9 per hour for her babysitting services. She babysits to earn money for her class trip. She needs to earn more than $135 for the trip. Which solution best represents the number of hours Nicole needs to babysit?

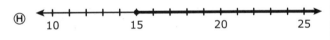

5 The water level in Ty's swimming pool has a maximum depth of 4 feet. After a day of swimming with friends, the water level in the pool drops 6 inches. Which of the following could best be used to determine d, the depth of the water in feet, after Ty and his friends went swimming?

Ⓐ $d - 0.5 \leq 4$

Ⓑ $d - 6 \leq 4$

Ⓒ $d + 0.5 \leq 4$

Ⓓ $d + 0.5 \geq 4$

3 The temperature dropped 11° in the last hour. The temperature is currently below 50°F. Which of the following could be used to determine the temperature, t, one hour ago?

Ⓐ $t - 11 = 50$

Ⓑ $t - 11 < 50$

Ⓒ $t - 11 > 50$

Ⓓ $t - 11 \leq 50$

6 Lakeisha and three friends order pizza and soda. They split the cost of the meal equally. If each girl spends $6.50 for the meal, which solution best represents the total cost of the meal?

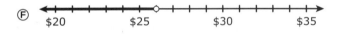

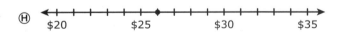
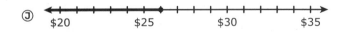

Name _____

Standards 6.9(A) – Supporting, 6.9(B) – Supporting

Unit 29 Independent Practice

1. Samantha must collect a minimum of 25 pounds of canned food in order to participate in the rodeo queen contest. If Samantha has already collected 13 pounds, which solution best represents the number of pounds of food she can collect and be able to participate in the contest?

   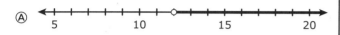

2. Meredith deposits $150 into her bank account. With the deposit, she now has over $500 in her account. Which of the following could be used to find x, the original balance before the deposit was made?

   Ⓕ $x - 150 > 500$

   Ⓖ $x + 150 \geq 500$

   Ⓗ $x - 150 < 500$

   Ⓙ $x + 150 > 500$

Use the information below to answer questions 2 and 3.

Mr. Milstein needs a larger kitchen table. The top of his current kitchen table measures $17\frac{1}{2}$ square feet. Mr. Milstein has decided to build a new table with a width of 5 feet.

2. Which of the following could be used to determine the possible length, l, of Mr. Milstein's new kitchen table?

   Ⓕ $5l > 17\frac{1}{2}$

   Ⓖ $2l + 2(5) > 17\frac{1}{2}$

   Ⓗ $l \cdot 5 \geq 17\frac{1}{2}$

   Ⓙ $17\frac{1}{2} - 5 > l$

Use the information below to answer questions 5 and 6.

Angelica earns $7 each week for her allowance. She saves $2.50 of her allowance each week for a pair of jeans that cost $45.

5. Which equation best represents the number of weeks, w, Angelica needs to save for the jeans?

   Ⓐ $\dfrac{w}{2.50} = 45$

   Ⓑ $2.50w = 45$

   Ⓒ $7w = 45$

   Ⓓ $45 - w = 2.50$

6. Which solution best represents the number of weeks Angelica must save before she has enough money to purchase the jeans?

3. Which number line best represents the possible lengths of Mr. Milstein's new table?

   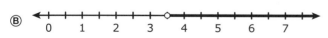

Unit 29 Assessment

Standards 6.9(A) – Supporting, 6.9(B) – Supporting

1. Betsy drew a pair of complementary angles. The first angle measured 43°. Which of the following could be used to determine the measure of s, the second angle Betsy drew?

   Ⓐ $s + 43 = 180$

   Ⓑ $43 + s = 90$

   Ⓒ $43 - s = 90$

   Ⓓ $180 - s = 43$

Use the information below to answer questions 2 and 3.

Mrs. Crimm signed up to bring two dozen cupcakes for the school bake sale. She made one dozen, but her children ate 3 of the cupcakes.

2. Which of the following could be used to determine c, the number of cupcakes Mrs. Crimm needs to bake?

   Ⓕ $24 - 3 = c$

   Ⓖ $c + 9 = 2$

   Ⓗ $9 + c \leq 24$

   Ⓙ $c + 9 = 24$

3. Which solution best represents the number of cupcakes Mrs. Crimm needs to bake?

   Ⓐ

   Ⓑ

   Ⓒ

   Ⓓ

Use the information below to answer questions 4 and 5.

Isabell and her three sisters plan to purchase a gift for their mother's birthday and split the cost equally. The total cost of the gift must be less than $60.

4. Which best represents c, the amount each sister might pay for their mother's birthday present?

   Ⓕ $3c < 60$

   Ⓖ $c \cdot 4 = 60$

   Ⓗ $4c < 60$

   Ⓙ $c \cdot 3 \leq 60$

5. Which number line best represents the amount of money each sister might pay for the gift?

   Ⓐ

   Ⓑ

   Ⓒ

   Ⓓ

6. Gaylon receives a $20 iTunes® gift card. Before receiving the gift card, his account had a balance of at least $2.50. Which solution best represents the amount of money Gaylon has in his account after receiving the gift card?

   Ⓕ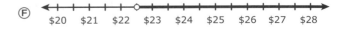

   Ⓖ

   Ⓗ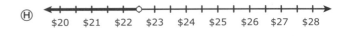

   Ⓙ

Name _____

Standards 6.9(A) – Supporting, 6.9(B) – Supporting

Unit 29 Critical Thinking

1. The students in Mr. Wells' math class write an inequality for the following situation and graph the solution on a number line.

Mrs. Taylor receives a maximum of 20 e-mails per day. Today, she has already received 8 e-mails. How many e-mails can she expect to receive for the remainder of the day?

Savannah and Parker each respond as shown below.

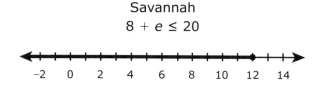
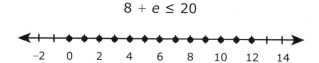

Savannah
$8 + e \leq 20$

Parker
$8 + e \leq 20$

Which student's response is correct? Explain.

_____
_____

2. What clues are used in determining whether a situation is represented by an equation or an inequality? Use the Venn diagram to compare and contrast situations that are represented by equations with those represented by inequalities.

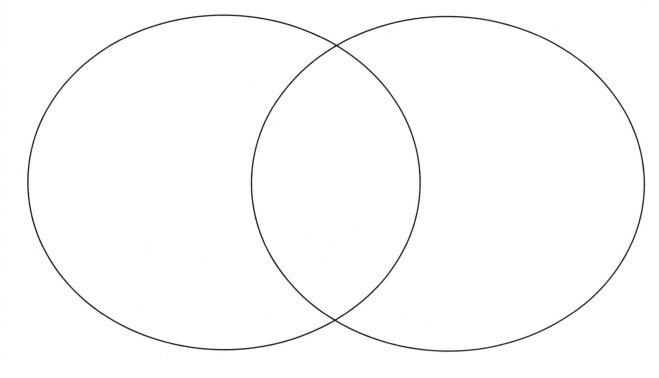

Unit 29 Journal/Vocabulary Activity

Standards 6.9(A) – Supporting, 6.9(B) – Supporting

Journal

Explain how to represent all solutions to $x \neq 3$. Use the number line to aid in your explanation.

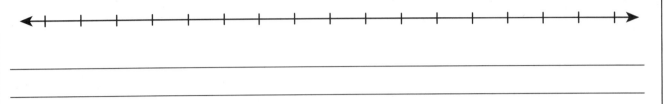

_____

_____

_____

Vocabulary Activity

Circle the correct responses to describe the representations of each situation.

1. Sheila rode the elevator up for more than 4 floors.
   (equation/inequality) (number line: shade right/shade left/no shade) (open circle/solid circle)

2. Yessica purchased 6 quarts of milk for $9.54.
   (equation/inequality) (number line: shade right/shade left/no shade) (open circle/solid circle)

3. The jar held a minimum of 15 lollipops.
   (equation/inequality) (number line: shade right/shade left/no shade) (open circle/solid circle)

4. Jeffery had at most $100 in his checking account.
   (equation/inequality) (number line: shade right/shade left/no shade) (open circle/solid circle)

5. Celine worked 32 hours last week and earned $232.
   (equation/inequality) (number line: shade right/shade left/no shade) (open circle/solid circle)

6. The tree had no more than 20 gifts under it.
   (equation/inequality) (number line: shade right/shade left/no shade) (open circle/solid circle)

7. The Vikings scored fewer than 15 touchdowns all season.
   (equation/inequality) (number line: shade right/shade left/no shade) (open circle/solid circle)

8. The minimum grade on the last math test was 64.
   (equation/inequality) (number line: shade right/shade left/no shade) (open circle/solid circle)

9. The temperature was less than 30°F when we woke up this morning.
   (equation/inequality) (number line: shade right/shade left/no shade) (open circle/solid circle)

10. The Heard family drove 855 miles in $13\frac{2}{13}$ hours to get to the beach.
    (equation/inequality) (number line: shade right/shade left/no shade) (open circle/solid circle)

When are 1500 plus 20 and 1600 minus 40 the same thing?

For each problem, write an equation or inequality that could be used to determine a solution. Find the equation or inequality in the box at the bottom of the page, and record its letter in the space next to the problem. When all the problems have been matched with an equation or inequality, use the letters to fill in the corresponding blanks below and reveal the answer to the title question.

__M__ __I__ __L__ __I__ __T__ __A__ __R__ __Y__   __T__ __I__ __M__ __E__
 3   1   8   1   4   2   7   5   4   1   3   6

1. __I__ The sum of Bernice's age and her mother's age is 63. Bernice is 21. What is her mother's age?

   Equation/Inequality: _____

2. __A__ The store ordered less than 500 cases of eggs for the holiday week. The eggs were equally distributed throughout the week. How many cases were distributed each day?

   Equation/Inequality: _____

3. __M__ Maura and Louis have $75 together to purchase a gift for their father. Maura has $40. How much money does Louis have?

   Equation/Inequality: _____

4. __T__ Jamal has saved over $2,000 for his vacation. He spends $642 on his plane ticket. How much does Jamal have to spend on the rest of his vacation?

   Equation/Inequality: _____

5. __Y__ The cost of one gallon of orange juice is $2.68. Michelle has $15. What is the maximum number of gallons of orange juice Michelle can buy?

   Equation/Inequality: _____

6. __E__ Kayla has 3 bottles of water. Together, she and Marcie have at least 12 bottles of water. How many bottles of water could Marcie have?

   Equation/Inequality: _____

7. __R__ David divides his allowance equally over 5 days. Each day he has $4 to spend. How much is David's allowance?

   Equation/Inequality: _____

8. __L__ Randi gave 6 pieces of candy to each of her friends. She gave a total of 30 pieces of candy. How many friends were given candy?

   Equation/Inequality: _____

E: $3 + x \geq 12$	I: $x + 21 = 63$	C: $7x > 500$	M: $x + 40 = 75$
L: $6x = 30$	D: $2.68x < 15$	R: $\frac{x}{5} = 4$	W: $x + 21 < 63$
T: $642 + x > 2{,}000$	A: $7x < 500$	N: $3 + x \leq 12$	Y: $2.68x \leq 15$

Unit 29 Homework

Standards 6.9(A) – Supporting, 6.9(B) – Supporting

1. Oliver has $500 in his savings account on the last day of school. He will use some of the money throughout the summer to pay for food and activities, but he wants to have $200 in his account on the first day of school. Oliver allows himself to withdraw $25 per week. How much money can Oliver spend during the summer?

   _____

   Represent w, the number of weeks Oliver can withdraw money from his savings and keep his balance at $200.

   _____

   Use the number line to represent the solution to the situation above.

   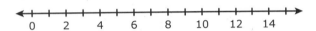

2. Frankie and Nick each graph the solution to the following inequality.

   $$8d \geq -40$$

   Frankie's solution

   Nick's solution

   Which student correctly graphed the solution to the inequality? Explain.

   _____
   _____
   _____

For problems 3–6, write an equation or inequality to represent the situation. Then graph the solution on the number line.

3. Sherry made a minimum of $75 babysitting last month, at a rate of $10 per hour. Number of hours Sherry babysat:

   _____

4. Kendra's grandmother uses $3\frac{3}{4}$ cups of sugar to make 2 batches of cookies. Cups of sugar per batch:

   _____

5. Larry scores a total of 142 points on two different math tests. He scores 70 on one test. Larry's score on the second test:

   _____

   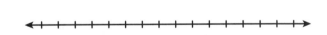

6. Glenda has less than $45 to spend on birthday gifts for her 5 friends. She spends an equal amount on each friend. The amount spent on one birthday gift:

   _____

   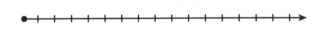

Connections

1. Use the Internet to research the word *inequality*. What does the word mean outside of math? How does this meaning relate to the math term? Write a paragraph about your findings to share with the class.

2. Divide a sheet of paper into four sections. Label the sections *Greater than*, *Less than*, *Greater than or equal to*, and *Less than or equal to*. Brainstorm phrases and words for each section of the paper.

Name _____

Standard 6.9(C) – Supporting

Unit 30 Introduction

For problems 1–5, represent each word problem using an equation or inequality.

1. Zane bought 2 packages of candy bars. He had fewer than 22 candy bars. Find b, the number of candy bars in each package.

2. Troy purchased 9 storage containers to help organize his garage. After beginning the project, he realized he would need 14 containers. What is c, the number of containers Troy still needs to purchase?

3. Mrs. Thompson measured 5 cups of caramel popcorn to fill each bag. She filled more than 25 bags with caramel popcorn. Find c, the total number of cups of popcorn Mrs. Thompson used to fill the bags.

4. Lena's brother loaned her $15. She now has at least $50 to purchase new boots. What is m, the amount of money Lena had before her brother loaned her the money?

5. Sidney received $5.21 in change after paying for her purchases with a $20 bill. Find c, the cost of Sidney's purchases.

For problems 6–8, create a real-world word problem that corresponds to the equation or inequality provided.

6. $3x = 42$

_____
_____
_____
_____
_____

7. $y + 12 < 25$

_____
_____
_____
_____
_____

8. $z - 2.5 \geq 10$

_____
_____
_____
_____
_____

Unit 30 Guided Practice

Standard 6.9(C) – Supporting

1 Which situation is best represented by the equation below?

$$x - 3.5 = 4.75$$

Ⓐ Alta had $4.75. She paid $3.50 for a snack and drink. What is x, the amount of money Alta has left?

Ⓑ Julia found a scarf with a sale price of $4.75. The discount was $3.50. What is x, the original price of the scarf?

Ⓒ Rocio drew a line segment that was 3.5 inches long. Renee drew a line segment that was 4.75 inches long. What is x, the difference in the lengths of the two line segments?

Ⓓ Ami completed a math problem in 3.5 minutes. Giovanni completed the same problem in 4.75 minutes. The time limit was 5 minutes. What is x, the difference in Ami's time and the time limit?

2 Which scenario is represented by the equation below?

$$16n = 48$$

Ⓕ Nate purchased 48 ounces of ground turkey. What is n, the number of pounds of ground turkey Nate purchased?

Ⓖ LaTisha bought 16 candy bars for 48¢ each. What is n, the total cost of the candy bars?

Ⓗ Savannah is 16 years old. Her father is 48 years old. What is n, the difference between Savannah's age and her father's age?

Ⓙ Derrick purchased 16 quarts of lemonade. He poured the lemonade into 48 cups. How many ounces, n, of lemonade were in each cup?

3 Which of the following could be represented by the inequality below?

$$y \cdot 12 \leq 144$$

I. The rectangle has a maximum area of 144 square inches. If the length is 12 inches, what is y, the width of the rectangle?

II. One movie ticket sells for $12. The Fisher family spent less than $144 on tickets for their family. Find y, the possible number of tickets the Fisher family purchased.

III. The florist received an order to assemble at most 144 red roses into small bouquets. Each bouquet will have a dozen red roses. How many bouquets, y, will the florist make?

Ⓐ I only   Ⓒ I and III only

Ⓑ I and II only   Ⓓ I, II, and III

4 Which situation could NOT be represented by the inequality below?

$$145 + z > 200$$

Ⓕ Kate buys a sled that costs more than $200. She pays $145 in cash and uses her credit card for the rest. Find z, the possible amount Kate charges to her credit card.

Ⓖ Missy had a collection of 145 seashells. On her trip to the beach, she found more seashells. Missy now has over 200 seashells in her collection. Find z, the number of seashells Missy found at the beach.

Ⓗ Charlie found the sum of two angles to be greater than 200°. If the measure of one angle is 145°, what is z, the measure of the second angle?

Ⓙ The regular price for a game system was at most $200. It was on sale this week for $145. Find z, the amount of money saved by purchasing the game system on sale.

Standard 6.9(C) – Supporting

Unit 30 Independent Practice

1 Which situation is represented by the equation below?

$$\frac{n}{45} = 5$$

Ⓐ Hannah was saving money so that she could buy a purse that cost $45. She only needs 5 more dollars. Find *n*, the amount Hannah has saved.

Ⓑ Dylan evenly distributed 45 baseball cards to his friends. Each friend received 5 cards. Find *n*, the number of friends that received baseball cards.

Ⓒ Kinsey must earn 45 badges in 5 months. She needs to earn *n* badges each month. Find the number of badges Kinsey needs to earn.

Ⓓ Marco made 45 bags of cookies. Each bag contained 5 cookies. Find *n*, the total number of cookies.

2 Which scenario is represented by the equation below?

$$c + 4 = 19$$

Ⓕ There are at least 4 boys in Mrs. Wright's first-period. She has 19 students in the class. What is *c*, the number of girls in Mrs. Wright's first period class?

Ⓖ Jamie's theater ticket cost $4 more than Kaley's theater ticket. If Jamie's ticket cost $19, what is *c*, the cost of Kaley's ticket?

Ⓗ Todd is four years younger than his sister, Claire. If Todd is 19 years old, what is *c*, Claire's age?

Ⓙ Hillary increased the length of a rectangle by 4 inches. If the perimeter is 19 inches, what is *c*, the original length of the rectangle?

3 Which situation is represented by the inequality below?

$$6x \geq 36$$

Ⓐ Brett built 6 squares using snap cubes. Each square was built with a minimum of 36 cubes. Find *x*, the total number of cubes Brett used.

Ⓑ There are at least 36 students waiting to enter the amusement park. They stand in 6 equal groups. What is *x*, the number of students in each group?

Ⓒ Fiona buys some earrings for $6 a pair. She pays at most $36 for all the earrings. Find *x*, the possible number of earrings Fiona purchased.

Ⓓ The area of a rectangle is greater than 36 square inches. If the width of the rectangle measures 6 inches, what is *x*, the length?

4 Which of the following could be represented by the inequality below?

$$x - 12.5 < 25$$

I. Mrs. Martin saved $12.50 by purchasing a pair of shoes on sale. She paid at most $25 for the shoes. What is *x*, the original price of the shoes?

II. Nicholas uses his debit card to purchase a CD for $12.50. After the transaction, Nicholas has a balance of less than $25 in his account. What is *x*, his initial balance?

III. When remodeling their home, Mr. Teasel decreased the area of his pantry by $12\frac{1}{2}$ square feet. The new pantry has an area smaller than 25 square feet. Find *x*, the original area of the pantry.

Ⓕ I only

Ⓖ II only

Ⓗ II and III only

Ⓙ I, II, and III

Unit 30 Assessment

Standard 6.9(C) – Supporting

1 Which situation is NOT represented by the equation below?

$$x - 5.7 = 3.25$$

Ⓐ Felix had a board that was x feet long. He removed 5.7 feet. Felix then had 3.25 feet left. What was the original length of the board?

Ⓑ Ginger received x dollars to go to the movies. Her ticket cost $5.70, and she had $3.25 left. How much money did Ginger take to the movies?

Ⓒ Anna sewed 5.7 meters of trim around a rug. She then sewed 3.25 meters of trim on a cushion. How much more trim, x, did Anna sew on the rug than the cushion?

Ⓓ Fabiola spent $5.70 of the x dollars she had in her wallet. She now has $3.25 in her wallet. What was the original amount of money Fabiola had in her wallet?

2 Which situation is best represented by the equation below?

$$c \cdot 3 = 2.1$$

Ⓕ Gavin gave the cashier $3 when the total was $2.10. Find c, the amount of change Gavin received.

Ⓖ Carley bought 3 packages of gum that each had the same cost. The total was $2.10. Find c, the cost of each package of gum.

Ⓗ Aunt Peggy made 2.1 pounds of fudge for each of her 3 nephews. Find c, the number of pounds of fudge Aunt Peggy made.

Ⓙ Chris must give a speech that is no longer than 3 minutes. He spoke for 2.1 minutes. Find c, the number of minutes Chris has left.

3 Which of the following can be represented by the inequality below?

$$\frac{g}{4} \leq 12$$

I. A square has a maximum perimeter of 12 centimeters. Find g, the possible side length of the square.

II. Zack purchased fewer than 12 donuts. If he split them evenly among his four best friends, what is g, the number of donuts each friend received?

III. The number of students in each group is at most 12. If there are 4 groups of students, what is g, the total number of students?

Ⓐ I only

Ⓑ I and II only

Ⓒ III only

Ⓓ I, II, and III

4 Which scenario is best represented by the inequality below?

$$3\frac{1}{2} + z > 10$$

Ⓕ Caleb cut $3\frac{1}{2}$ feet off a length of rope that is a minimum of 10 feet long. Find z, the amount of rope that is left.

Ⓖ The Walkers' road trip will take over 10 hours. They have traveled 3.5 hours so far. How many hours, z, do the Walkers have left to travel?

Ⓗ After she was born, Elliott gained $3\frac{1}{2}$ pounds and now weighs 10 pounds. What was z, Elliott's birth weight?

Ⓙ Juan is $3\frac{1}{2}$ years older than his sister. His sister is at least 10 years old. Find z, Juan's age.

Name _____

Standard 6.9(C) – Supporting

Unit 30 Critical Thinking

1 Donald completes the following problem on a math test.

Write a real-world problem to match the equation 8x = 104.

His response is shown below.

Jessica has collected 104 stuffed animals. She must pack them in boxes before her family moves to a new house. If she puts at least 8 animals in each box, how many boxes will Jessica need?

Donald's teacher marked his response incorrect. What mistake did Donald make?
_____
_____

Rewrite Donald's response so he answers the question correctly.
_____
_____

2 Given the following, write one word problem to match the equation and one word problem to match the inequality.

$$\frac{x}{14} = 112 \qquad\qquad \frac{x}{14} < 112$$

_____   _____
_____   _____
_____   _____
_____   _____
_____   _____
_____   _____

Which problem did you find easier to write, the equation word problem or the inequality word problem? _____

Explain why.
_____
_____

Unit 30 Journal/Vocabulary Activity

Name _____

Standard 6.9(C) – Supporting

Journal

A fellow student was absent during the lesson on equations and inequalities. Explain to the student the difference between writing a problem when given an equation and writing a problem when given an inequality.

_____
_____
_____
_____

Vocabulary Activity

For each symbolic representation shown below, state the symbol's meaning and then list words and/or phrases that would be represented by the symbol in a word problem.

$<$ Meaning: _____

Words/Phrases: _____
_____
_____

$>$ Meaning: _____

Words/Phrases: _____
_____
_____

$=$ Meaning: _____

Words/Phrases: _____
_____
_____

$\leq$ Meaning: _____

Words/Phrases: _____
_____
_____

$\geq$ Meaning: _____

Words/Phrases: _____
_____
_____

Name _____

Standard 6.9(C) – Supporting

Unit 30 Motivation Station

Guess My Inequality

Play *Guess My Inequality* with a partner. Each player needs a game sheet, a blank sheet of paper, and a pencil. Player 1 begins by selecting an inequality from the list provided. He/she records the inequality on the sheet of paper, being careful not to allow player 2 to see what is written. Player 2 begins by asking the questions listed, in order, and recording the responses on his/her sheet of paper. As soon as player 2 feels he/she can correctly write the inequality recorded by player 1, the inequality is recorded in the space provided below. If correct, player 2 earns the same number of points as the number of the last question asked. If incorrect, player 2 loses a turn, and the roles are reversed for the next round using a new inequality. The game ends after each player has guessed three rounds. The winner is the player with more points.

Inequalities

$14a > 72$ $m - 18 \leq 28$ $\frac{x}{3} < 12$ $9k < 63$

$10 + y \geq 25$ $b + 6 > 14$ $-5n \leq 25$ $c - 8 > 10$

Questions

START   10. Does your inequality use division to solve?

9. Is the solution to your inequality greater than or equal to a number?

8. Does your inequality involve addition?

7. Is the solution to your inequality less than a number?

6. Does your inequality use addition to solve?

5. Is the solution to your inequality greater than a number?

4. Does your inequality involve division?

3. Is the solution to your inequality less than or equal to a number?

2. Does your inequality use subtraction to solve?

1. Is the solution to your inequality equal to a number?

Guesses

Inequality	Points	Inequality	Points
_____	_____	_____	_____
_____	_____	_____	_____
_____	_____	_____	_____

Unit 30 Homework

Standard 6.9(C) – Supporting

1 Write a word problem to match each of the following equations.

$$7x = 35$$

_____
_____
_____
_____

$$7 + x = 35$$

_____
_____
_____
_____

$$x - 7 = 35$$

_____
_____
_____
_____

$$\frac{x}{7} = 35$$

_____
_____
_____
_____

2 Mr. Ellis gave his class the following inequality and asked them to write a scenario that could be solved using the inequality.

$$n - 14 \leq 18$$

Jillian wrote the following scenario:

Mike needs money to buy a part for his bike. He knows his dad will pay him $14 if he cleans out the garage. After that, he will need at most $18 more to buy the part. How much is the part for Mike's bike?

Elliott wrote the following scenario:

Jan and Kay are purchasing buddy passes to the carnival. Kay has $14 of the total amount they need to purchase the passes. With Jan's money, the two have less than $18. How much money does Jan have?

Which student wrote a correct scenario? Explain your answer.

_____
_____
_____
_____

Based on your response above, rewrite the incorrect scenario to make it correct.

_____
_____
_____
_____
_____
_____

Connections

Write a word problem that can be solved using a one-step equation or inequality. Ask a family member to write the equation or inequality that could be used to solve the problem. Explain how to solve the problem if they answer incorrectly.

Name _____

Standard 6.10(A) – Readiness

Unit 31 Introduction

1. Raymond is 4 years old. His sister is 3 years older. The equation represented on the balance scale can be used to determine *x*, the age of Raymond's sister.

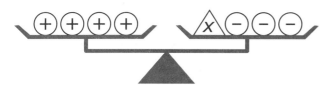

Write the equation to determine the age of Raymond's sister.

_____

Explain the process for solving the equation.

How old is Raymond's sister? _____

2. Sierra bought 3 snow cones at the spring carnival. She paid a total of $9.75. Write an equation that can be used to determine *x*, the price Sierra paid for each snow cone.

_____

Sketch a model of the equation. (Hint: Use bills and coins.)

Explain how you would use the model to determine the value of *x*.

What is the price of one snow cone?

_____

3. Mrs. Apparicio wants to purchase a painting for her living room. The painting can have a maximum area of 14 square feet. Mrs. Apparicio chooses a painting with a width of $3\frac{1}{2}$ feet. Write an equation or inequality that can be used to determine *l*, the possible length of the painting.

_____

What is the solution to the equation or inequality written above?

_____

4. Solve each inequality.

$\dfrac{x}{-7} \leq 17$ $\qquad$ $\dfrac{x}{7} \leq -17$

Explain the differences in solving the two inequalities.

_____
_____
_____

5. Explain, in words, the steps used to solve the following:

$$-16.3 > x - 3.4$$

_____
_____
_____

What is the solution to the inequality above?

_____

Unit 31 Guided Practice

Standard 6.10(A) – Readiness

1 Sam earns $4 per hour completing chores on the weekends. Last weekend, he earned $10. Use the model below to determine *x*, the number of hours Sam worked last weekend.

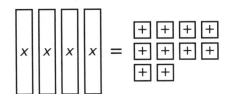

Ⓐ $x = 10$

Ⓑ $x = 2$

Ⓒ $x = 2\frac{1}{2}$

Ⓓ $x = 6$

Use the model to answer questions 2 and 3.

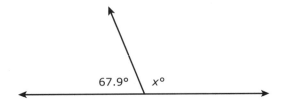

2 Carrie needs to determine the missing angle measure. Which of the following equations could Carrie use to determine *x*?

Ⓕ $360 - x = 67.9$

Ⓖ $67.9 + x = 180$

Ⓗ $180 + x = 67.9$

Ⓙ $67.9 - x = 180$

3 Which best represents the solution to Carrie's equation?

Ⓐ $x = 22.1°$

Ⓑ $x = 50.1°$

Ⓒ $x = 122.1°$

Ⓓ $x = 112.1°$

4 Marissa solves the inequality as shown below.

$$\frac{x}{-12} > 7.8$$

$$(-12)\frac{x}{-12} > 7.8(-12)$$

$$x > -93.6$$

Marissa's teacher checks her work and notices a mistake in her answer. What is Marissa's mistake?

Ⓕ Marissa should add 12 to both sides of the inequality.

Ⓖ Marissa incorrectly multiplies 7.8(-12).

Ⓗ Marissa's teacher could not read her work clearly, and her answer is actually correct.

Ⓙ Marissa did not change the direction of the inequality symbol after multiplying both sides of the inequality by a negative number.

5 Mrs. Mitchell purchases milk for $3.50 per gallon. Which represents the number of gallons of milk, *g*, Mrs. Mitchell can purchase if she plans to spend no more than $10.50 on milk?

Ⓐ $g \geq 0$ and $g \leq 3$

Ⓑ $g \geq 0$ and $g \leq 14$

Ⓒ $g \geq 0$ and $g \leq 7$

Ⓓ $g \geq 0$ and $g \leq 2$

Standard 6.10(A) – Readiness

Unit 31 Independent Practice

1. Kwan uses a balance scale to solve an equation, as shown below.

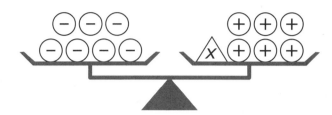

Which value of x best represents the solution to the equation?

Ⓐ $x = -13$

Ⓑ $x = -1$

Ⓒ $x = 13$

Ⓓ $x = 6$

2. Mrs. Bryan saves $1,500 for a family vacation. She budgets $235 of her savings for transportation. Which of the following represents the equation and solution to determine x, the amount of money Mrs. Bryan can budget for the rest of the vacation?

Ⓕ $x - 235 = 1,500; x = 1,735$

Ⓖ $235 + x = 1,500; x = 1,265$

Ⓗ $1,500 + 235 = x; x = 1,735$

Ⓙ $x - 1,500 = 235; x = 1,265$

3. Millie and her brother, Jack, sold candy bars for the band fund-raiser. Millie sold three times as many candy bars as Jack. If Jack sold 30 candy bars, which equation and solution can be used to represent y, the number of candy bars Millie sold?

Ⓐ $3y = 30; y = 10$

Ⓑ $30 = \frac{y}{3}; y = 103$

Ⓒ $30 = \frac{y}{3}; y = 90$

Ⓓ $3y = 30; y = 90$

4. Sam calls a friend to get help with his math homework. Which instructions would best help Sam solve the inequality shown below?

$$15\frac{3}{4} < -2 + x$$

Ⓕ Add $15\frac{3}{4}$ to both sides of the inequality.

Ⓖ Subtract 2 from both sides of the inequality.

Ⓗ Add 2 to both sides of the inequality.

Ⓙ Subtract $15\frac{3}{4}$ from both sides of the inequality.

5. Leanne drew two angles. The sum of the measures of the two angles cannot be more than 60°.

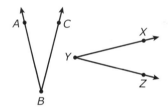

Leanne determines the measure of ∠XYZ to be 27.8°. Which best describes b, the possible measures of ∠ABC?

Ⓐ $b > 0$ and $b \leq 87.8$

Ⓑ $b > 0$ and $b \leq 32.2$

Ⓒ $b > 0$ and $b < 87.8$

Ⓓ $b > 0$ and $b < 32.2$

6. Which best represents the solution for the inequality shown below?

$$-100 > -5y$$

Ⓕ $95 > y$

Ⓖ $20 > y$

Ⓗ $95 < y$

Ⓙ $20 < y$

Unit 31 Assessment

Standard 6.10(A) – Readiness

1. Mrs. Magill asks her students to draw a pair of complementary angles with one angle measuring 63°. Which equation and solution process can be used to determine a, the measure of the other angle Mrs. Magill's students should draw?

   A) $63 + a = 180$; subtract 63 from both sides of the equation

   B) $a - 63 = 90$; add 63 to both sides of the equation

   C) $63 + a = 90$; subtract 63 from both sides of the equation

   D) $a - 63 = 180$; add 63 to both sides of the equation

2. Which best represents the correct way to solve the inequality shown below?

   $$-8.2 \geq 4x$$

   F) $\frac{-8.2}{-4} \geq \frac{4x}{-4}$

   $2.05 \leq x$

   G) $\frac{-8.2}{4} \geq \frac{4x}{4}$

   $-2.05 \geq x$

   H) $\frac{-8.2}{4} \geq \frac{4x}{4}$

   $-2.5 \geq x$

   J) $\frac{-8.2}{4} \geq \frac{4x}{4}$

   $-2.05 \leq x$

3. Evelyn plans to go to the fair with three of her friends. She needs to earn at least $60 to attend the fair. If Evelyn earns $7.50 per hour babysitting, how many hours will she need to work in order to have enough money to go to the fair with her friends?

   A) $x < 8$

   B) $x \leq 8$

   C) $x > 8$

   D) $x \geq 8$

4. Sawyer and his brother evenly split the cost of dinner. The total cost for dinner was $15. Based on the model below, what is x, the amount of money Sawyer paid for dinner?

   F) $15.00

   G) $7.00

   H) $5.00

   J) $7.50

5. Mr. Sims received $3.36 in change from his grocery purchase. If the total amount of his purchase was $56.64, which equation and solution best represent x, the amount of money Mr. Sims gave the cashier?

   A) $x - 56.64 = 3.36$; $x = 60$

   B) $x + 3.36 = 56.64$; $x = 53.28$

   C) $56.64 - x = 3.36$; $x = 60$

   D) $x + 56.64 = 3.36$; $x = 60$

Name _____

Standard 6.10(A) – Readiness

Unit 31 Critical Thinking

1 One prism balances four cubes, and one cylinder balances five cubes. The relationships are shown in the illustrations below.

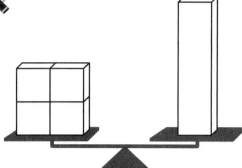 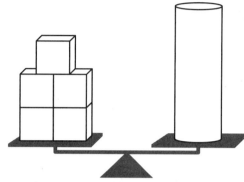

Based on the relationships shown, determine whether each of the following statements is *true* or *false*. Justify your answers using words or pictures.

a. One prism and one cube balance five cubes. _____

b. One prism and one cube balance one cylinder. _____

c. One cylinder and one prism balance ten cubes. _____

2 The number line represents a solution set for an inequality.

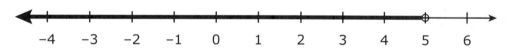

Use the following descriptions to write inequalities that have the same solution set as represented by the number line above. Justify each answer by solving the inequality.

a. One-step, use subtraction to solve

b. One-step, use division to solve

c. One-step, use division by a negative number to solve

Name _____

 Unit 31 Journal/Vocabulary Activity

Standard 6.10(A) – Readiness

Journal

Explain how using multiplication or division when solving an equation is different from using multiplication or division when solving an inequality.

_____
_____
_____
_____

Vocabulary Activity

Match each description to a number sentence from the box. Record the letter from the number sentence in the blank before the description. Each answer will be used only once.

_____ 1. Shows two *equivalent expressions*

_____ 2. Has one *term* with a *coefficient* of -8

_____ 3. The *inverse operation* used to solve is addition.

_____ 4. Contains one *variable* and one *constant*

_____ 5. The *expression* on the right of the *inequality* symbol contains a *variable*.

_____ 6. The *solution* to the *inequality* is $x \leq -4$.

A	$x = -8$
B	$-8x$
C	$-3x \geq 12$
D	$7 + x = 19$
E	$14 < x + 6$
F	$x - 4 \geq -8$

For each of the following terms, draw a pictorial example. Label the example with values.

7. *Complementary angles*

8. *Supplementary angles*

9. *Triangle inequality theorem*

10. *Linear angle pairs*

Name _____

Standard 6.10(A) – Readiness

Unit 31 Motivation Station

A Maze of Solutions

Beginning at START on the maze, find the solution to the model, equation, or inequality and use the solution to shade the path to follow. The correct path leads to the FINISH of the maze.

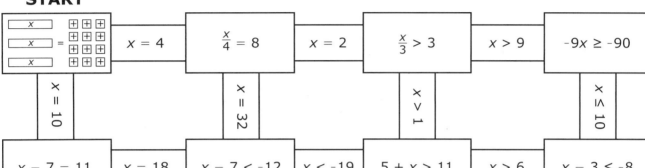

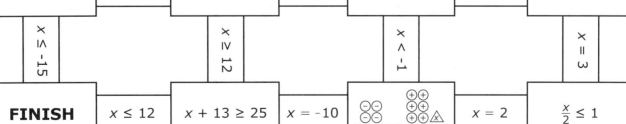

©2014 mentoringminds.com · motivationmath™ LEVEL 6 · ILLEGAL TO COPY · 253

Unit 31 Homework

Standard 6.10(A) – Readiness

1 Explain, in words, the steps used to solve the following:

$$-2.75b \leq 33$$

_____

_____

_____

What is the solution to the inequality above?

_____

2 Tisha draws a triangle for her math class, using the following information:

- One side of the triangle must measure 3 inches.
- One side of the triangle must measure $1\frac{3}{4}$ inches.
- The third side should be an appropriate length to form a triangle.

Write 3 inequalities that can be used to find the possible length, x, of the third side of Tisha's triangle.

_____

_____

_____

Using the inequalities written above, what is the range of possible side lengths for the triangle Tisha draws?

Justify your answer.

3 The cost to enter a state park for one day is $5 per person over the age of 12. Children under age 12 enter for free. On one day in July, Cooper Lake State Park collected $1,380. Write an equation that can be used to find p, the number of people over the age of 12 that entered the park on that one day.

How many people over the age of 12 entered Cooper Lake State Park on one day in July?

4 Write an equation to represent the model.

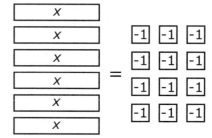

What value of x makes the equation true?

5 At Glenn Middle School, 535 students chose football as their favorite sport to watch. That is 58 less than the number of students who chose basketball. Write and solve an equation to find b, the number of students who chose basketball as their favorite sport to watch.

Connections

Research the current exchange rate for U.S. money in 2 different foreign countries. Write equations that can be used to convert U.S. money to the type of money used in each of the countries you chose. Using the amount of money you have, determine how much money you would have in the other countries. In which country is your U.S. money worth more? Why do you think that is true? Record your findings and share with the class.

Standard 6.10(B) – Supporting

Unit 32 Introduction

1. Mia, Nancy, Olivia, and Paul must determine values that are true for this inequality:

    $$\frac{x}{8} > 3$$

    The table below shows their responses.

Student	x
Mia	30
Nancy	22
Olivia	24
Paul	32

    Which student(s) correctly determined a value for x?

    _____

    Explain why all the values in the table are not true for the inequality.

    _____

    _____

2. Callaway's has a sale on pocket tissue this week for $0.85 each. Tracee spends less than $6 on tissue. She uses the inequality below to determine how many pocket tissues she can purchase.

    $$0.85p < 6$$

    Can Tracee purchase 9 pocket tissues from Callaway's? Explain why or why not.

    _____

    _____

    List all possible values of p that are true for the number of pocket tissues Tracee can purchase.

    _____

3. For which of the following is $x = -3$ a solution? Write *true* or *false* in the blanks provided.

    - $-12 + x = -15$ _____

    - $-24 > -15x$ _____

    - $x - 5\frac{2}{3} \geq -9$ _____

    - $\frac{x}{1.25} > -2.4$ _____

    - $-\frac{6}{5} = x - \frac{3}{5}$ _____

4. Four students were asked to solve this equation:

    $$18.33 = 3.9y$$

    Each of the four students found a different solution for y, as shown in the table below.

Student	Solution
Kenley	$y = 14$
Garrison	$y = 4.7$
Sydney	$y = 71.487$
Mario	$y = 47$

    Which student(s) determined a true solution for the equation provided?

    _____

    Explain.

    _____

    _____

Unit 32 Guided Practice

Standard 6.10(B) – Supporting

1. Mr. Montoya purchases one dozen roses. He pays $3.50 per rose. He uses this equation to determine the cost of the roses:

$$\frac{c}{12} = 3.50$$

Which value represents the cost of Mr. Montoya's purchase?

- Ⓐ $105
- Ⓑ $42
- Ⓒ $32
- Ⓓ $41

2. Mr. Warren has 22 quarters in his truck. This is less than four times the number of quarters Mrs. Warren has in her van. The inequality shown below expresses the relationship between the numbers of quarters they have.

$$4n > 22$$

Which of the following could be the number of quarters Mrs. Warren has in her van?

- Ⓕ 4
- Ⓖ 5
- Ⓗ 5.5
- Ⓙ 6

3. Laurie rode her bike 17.5 miles in 2 hours. She uses this equation to represent the relationship between her time and distance.

$$2x = 17.5$$

What does the solution $x = 8.75$ represent?

- Ⓐ The distance Laurie rode her bike
- Ⓑ The time Laurie spent riding her bike
- Ⓒ The distance Laurie traveled per hour
- Ⓓ Not here

4. Which represents a possible solution set for the inequality shown below?

$$\frac{x}{-3} < 10$$

- Ⓕ {-60, -9, -3, 15}
- Ⓖ {-25, -10, 12, 27}
- Ⓗ {-30, 3, 30, 60}
- Ⓙ {-40, -38, -36, -34}

5. Jayden has completed $3\frac{1}{2}$ pages of his social studies report. Marshall has completed $7\frac{1}{4}$ pages of the same report. They use the equation below to represent their current progress on the report.

$$3\frac{1}{2} + p = 7\frac{1}{4}$$

Which best describes p in the equation?

- Ⓐ $3\frac{3}{4}$, the difference between the number of pages Jayden and Marshall have completed
- Ⓑ $10\frac{3}{4}$, the total number of pages Jayden and Marshall have completed
- Ⓒ $3\frac{3}{4}$, the number of pages Jayden still needs to complete
- Ⓓ $10\frac{3}{4}$, the number of pages the social studies report needs to be

6. Tamara saves a minimum of $250 for school clothes. She spends $75 of her savings on a pair of shoes. The inequality below can be used to determine the amount of money Tamara still has to spend.

$$250 \leq x + 75$$

What value of x makes the inequality true?

- Ⓕ 150
- Ⓖ 160
- Ⓗ 180
- Ⓙ 170

Standard 6.10(B) – Supporting

Unit 32 Independent Practice

1 Ms. Oliver sells bookmarks for $0.96 each to raise money to purchase new library books. So far, she has collected a total of $69.12. Ms. Oliver uses the equation below to represent her bookmark sales.

$$0.96b = 69.12$$

Which of the following could represent the number of bookmarks Ms. Oliver has sold so far?

Ⓐ 72 bookmarks

Ⓑ 69 bookmarks

Ⓒ 68 bookmarks

Ⓓ 7 bookmarks

2 Tabitha is 9 years old. This is less than half of Ariel's age, a. Tabitha and Ariel use the following inequality to represent the relationship between their ages:

$$9 < \frac{a}{2}$$

Which of the following is NOT a possible value for a?

Ⓕ 21   Ⓗ 20

Ⓖ 17   Ⓙ 19

3 Justin needed to cook 125 hot dogs for the tailgate party. His grill ran out of gas after cooking only 38 hot dogs. Justin uses the equation below to determine how many hot dogs he still needs to cook.

$$125 = 38 + h$$

Which value represents the number of hot dogs Justin still needs to cook?

Ⓐ 97   Ⓒ 87

Ⓑ 113   Ⓓ 63

4 Which equation is NOT true for the value of $x = 2\frac{1}{2}$?

Ⓕ $x + 3 = 5.5$

Ⓖ $2x = 5$

Ⓗ $x \div 2 = 1.25$

Ⓙ $x - 10 = 7.5$

5 Liv needs more than 36 people to sign her petition. She collects 15 signatures at school and around her neighborhood. Which statement is true about the inequality $15 + s > 36$?

Ⓐ Liv still needs fewer than 20 signatures.

Ⓑ Liv still needs at most 20 signatures.

Ⓒ Liv still needs more than 21 signatures.

Ⓓ Liv still needs at least 21 signatures.

6 All the items at Phillip's garage sale are $0.75. His goal is to earn at least $50. Using the inequality, $50 \leq 0.75x$, which represents a number of items Phillip can sell and meet his goal?

Ⓕ 62

Ⓖ 64

Ⓗ 66

Ⓙ 68

Unit 32 Assessment

Standard 6.10(B) – Supporting

1 Javier scored 214.8 points playing the first round of a video game. His sister played the second round and scored 198.4 points. Using the equation $x - 214.8 = 198.4$, which of the following best represents the total points scored after the second round?

Ⓐ 16.4 points

Ⓑ 302.2 points

Ⓒ 313.2 points

Ⓓ 413.2 points

2 Roberto saved at most $810 from his job washing cars during the summer. He plans to use $30 per week during the school year for meals and spending money. He uses the inequality $30w \leq 810$ where w is the number of weeks. Which of the following could NOT represent the number of weeks Roberto's money will last?

Ⓕ 23

Ⓖ 29

Ⓗ 27

Ⓙ 25

3 Yuri purchases a bookcase for her room. Each shelf contains 13 books which is $\frac{1}{4}$ the total number of books Yuri owns. Which of the following represents the number of books Yuri owns based on the equation, $\frac{1}{4}b = 13$?

Ⓐ $b = 52$

Ⓑ $b = \frac{13}{4}$

Ⓒ $b = \frac{4}{13}$

Ⓓ $b = 42$

4 For which of the following is 6.3 NOT a possible solution?

Ⓕ $5x > 30$

Ⓖ $12 \geq x - 2.7$

Ⓗ $\frac{x}{3.2} \leq 12$

Ⓙ $40.9 > 34.7 + x$

5 Quintavion builds a flower box for a local charity. The flower box is square shaped with side lengths of $5\frac{3}{4}$ feet. Quintavion uses the equation below to express the relationship between the side lengths and the number of sides.

$$\frac{y}{4} = 5\frac{3}{4}$$

Which best describes y in the equation?

Ⓐ $1\frac{7}{16}$, the perimeter, in feet, of the square

Ⓑ $20\frac{3}{4}$, the area, in square feet, of the flower box

Ⓒ 23, the perimeter, in feet, of the square

Ⓓ $9\frac{3}{4}$, the area, in square feet, of the flower box

6 Jerome must read over 300 pages this grading period for his English class. He uses the inequality below to show the relationship between the number of pages he must read and the number of pages he has already completed.

$$300 < 26 + p$$

Which of the following could represent p, the number of pages Jerome still needs to read?

Ⓕ 274

Ⓗ 270

Ⓖ 276

Ⓙ Not here

Name _____

Standard 6.10(B) – Supporting

Unit 32 Critical Thinking

1 Tamecia's parents limit the number of text messages she is allowed to send each month to at most 1,000. So far this month, Tamecia has sent 642 messages. Write an inequality that can be used to determine m, the number of messages Tamecia could send for the remainder of the month.

Inequality: _____

For each of the following, indicate whether the value is a reasonable solution to the inequality for the situation by writing *yes* or *no* in the space provided. Explain your responses.

400 _____

-3 _____

300 _____

350.5 _____

358 _____

2 Write an equation for each of the solutions given, using the operation shown.

$x = -12$; multiplication $x = 8$; addition $x = \frac{3}{2}$; division

_____   _____   _____

Justify each equation is correct by showing each solution below.

Unit 32 Journal/Vocabulary Activity

Standard 6.10(B) – Supporting

Journal

Mr. Fox displays the following statement:

For any given linear equation, there exists one and only one value for x that makes the equation true.

Explain Mr. Fox's statement.

_____
_____
_____

Vocabulary Activity

For each situation, determine if the solution to the equation or inequality written is *finite* or *infinite*, and *discrete* or *continuous*. Record responses by circling the correct terms.

1. Jameson has more than 5 whole cookies. He shares 2 whole cookies with Billy. Write an inequality to find c, the number of whole cookies Jameson could still have for himself.

    finite/infinite   discrete/continuous

2. After a decrease of 5°, the current temperature is 25°F. Write an equation to find t, the temperature before the decrease.

    finite/infinite   discrete/continuous

3. A hot air balloon flies at a speed of 10 miles per hour. Write an inequality to find h, the number of hours it could take a balloon to travel at least 70 miles.

    finite/infinite   discrete/continuous

4. At the hardware store, a minimum purchase of 8 ounces of nails is required. Write an inequality to find p, the price of purchasing nails if the cost per ounce is $0.05.

    finite/infinite   discrete/continuous

5. John uses 27 feet of wire for a sculpture in art class. Write an equation to find y, the number of yards of wire John uses.

    finite/infinite   discrete/continuous

6. Nancy needs no more than 35 beads to make a friendship bracelet. She already has 12 beads. Write an inequality to find b, the number of beads Nancy could use to make a bracelet.

    finite/infinite   discrete/continuous

Name _____

Standard 6.10(B) – Supporting

Unit 32 Motivation Station

All or Nothing! (or Some)

Play *All or Nothing! (or Some)* with a partner. Each pair of players needs a number cube, a game board, and a paper clip to use with the spinner. Each player needs a pencil. Player 1 spins the spinner to determine an equation or inequality, and rolls the number cube to determine a number set from the table. For the equation/inequality spun, the player determines whether all of the elements in the number set are true, some of the elements are true, or none of the elements are true. Player 1 records the equation or inequality and the elements from the number set that are true, if applicable, in the table. If none of the elements are true for the equation/inequality, player 1 writes *None* in the space. The player earns the number of points equal to the sum of the elements that are true. Play passes to player 2 who repeats the process. The game ends after 6 rounds. The winner is the player with more points.

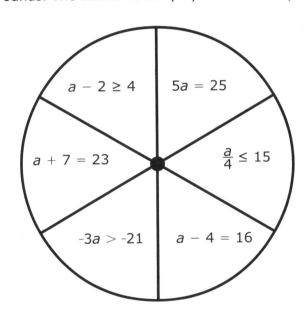

1	{-3, 0, 3}	4	{7, 16, 20}
2	{6, 15, 24}	5	{20, 30, 40}
3	{5, 15, 25}	6	{8, 12, 16}

Player 1

Equation/Inequality	True Elements	Points
TOTAL POINTS		

Player 2

Equation/Inequality	True Elements	Points
TOTAL POINTS		

Unit 32 Homework

Name _____

Standard 6.10(B) – Supporting

1 For each inequality, match the solution set that best represents possible solutions. Write the letter of the set next to the inequality.

_____ $b - 7 < -12$ **A** {-2, 0, 2}

_____ $8b > 40$ **B** {8, 9, 10}

_____ $-1 + b \geq 4$ **C** {-5, -7, -9}

_____ $\frac{b}{-5} \geq 1$ **D** {-6, -8, -10}

_____ $2 + b > -3$ **E** {-4, -2, 0}

_____ $-12b > -60$ **F** {5, 6, 7}

2 For which equations is $x = -2$ true? Write *true* or *false* in the blanks.

a. $9x = 18$ _____

b. $x + 4 = 2$ _____

c. $x - 7 = -9$ _____

d. $\frac{x}{-4} = 8$ _____

3 Explain the differences between the solution sets for each of the following:

$6x = 40$ $6x < 40$ $6x \leq 40$

_____
_____
_____

4 Jana invites fewer than 12 friends to her slumber party. Three friends have accepted her invitation. Write an inequality that can be used to find *f*, the number of friends that have not yet accepted Jana's invitation.

_____

What is the solution set for the situation above?

_____

5 Four students were asked to write an equation with a solution of 5. The students wrote the following:

- Allison: $8x = 13$
- Jennifer: $-4 + x = 1$
- Cathy: $x - 3 = 2$
- Karen: $\frac{x}{-1} = -5$

Which students wrote a correct equation? Justify your answer.

_____
_____
_____

6 For the inequality $6 + a > 12$, list two solutions and two non-solutions. Justify your answer.

Solutions: _____

_____

Non-solutions: _____

_____

Connections

Write one-step equations and inequalities on index cards, one per card. Place the cards face down between all players. Using 3–5 number cubes, one person rolls the cubes and turns over the top card. That person determines whether the numbers rolled are solutions to the equation or inequality on the card. If the answer is correct, the player keeps the card. If incorrect, the card is placed face down on the bottom of the pile. Play continues until all cards are used. The winner is the person with the most cards.

Standard 6.11(A) – Readiness

Unit 33 Introduction

Use the coordinate grid to answer questions 1–9.

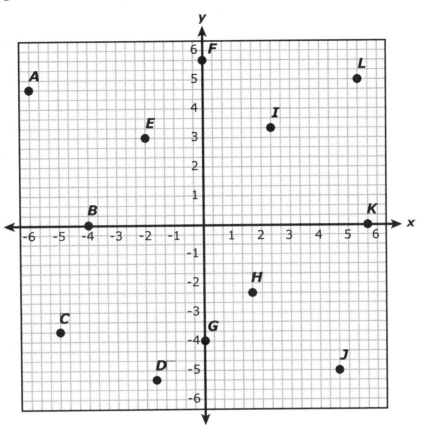

1. Which points are located in Quadrant IV?

2. Which point is located at (-4, 0)?

3. What are the coordinates for point D?

4. Which points have a y-value greater than 4?

5. Which points are located in Quadrant II?

6. Which points have an x-value of 0?

7. What is the ordered pair that is located 2 whole units to the right and 3 whole units down from point E?

8. Which points have negative values for both x and y?

9. If points F and I are connected to form a line segment, name another ordered pair on the line segment.

Unit 33 Guided Practice

Standard 6.11(A) – Readiness

Use the polygon shown on the coordinate grid to answer questions 1–3.

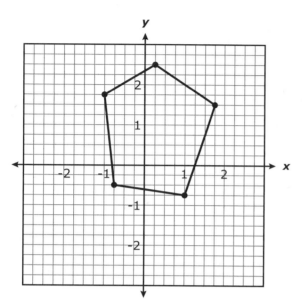

1. Which ordered pair does NOT represent a vertex of the polygon?

   A) $\left(-\frac{3}{4}, -\frac{1}{2}\right)$

   B) $\left(\frac{1}{4}, 2\frac{1}{2}\right)$

   C) $\left(-1\frac{1}{2}, 1\frac{3}{4}\right)$

   D) $\left(1, -\frac{3}{4}\right)$

2. Which ordered pair lies inside the polygon and is located in Quadrant IV?

   F) $\left(-\frac{1}{2}, -\frac{3}{4}\right)$

   G) $\left(\frac{3}{4}, -\frac{1}{2}\right)$

   H) $\left(-\frac{1}{4}, -\frac{1}{2}\right)$

   J) $\left(\frac{1}{2}, -\frac{3}{4}\right)$

3. Which point is located on the perimeter of the polygon?

   A) $\left(-\frac{1}{2}, 2\frac{1}{4}\right)$

   B) $\left(1, -\frac{1}{2}\right)$

   C) $\left(\frac{3}{4}, 2\frac{1}{4}\right)$

   D) $\left(1\frac{1}{2}, \frac{3}{4}\right)$

Use the map to answer questions 4 and 5.

The routes Tia takes from her house to different places are represented on the grid below.

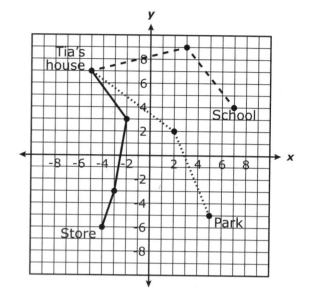

4. Which ordered pair best represents a point on Tia's route to the store?

   F) (-5, 6)

   G) (-2.5, 0)

   H) (-2, 5)

   J) (-3.5, -3)

5. Each unit on the grid represents 1 mile. For Tia to travel from the park to the library, she must go 3 miles south and 5 miles west. Which represents the coordinates of the library?

   A) (0, -8)

   B) (10, -8)

   C) (-8, 0)

   D) (-8, 10)

Standard 6.11(A) – Readiness

Unit 33 Independent Practice

Use the grid to answer questions 1–3.

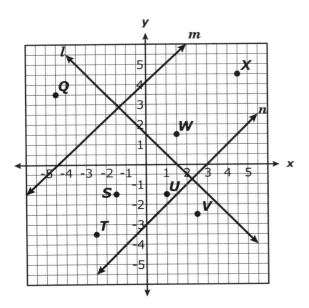

Use the grid to answer questions 4 and 5.

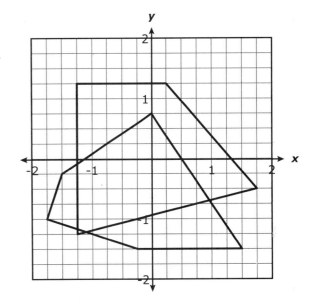

1 Which of the following are NOT coordinates located on line *n*?

- Ⓐ $\left(4\tfrac{1}{2},\ 1\tfrac{1}{2}\right)$
- Ⓑ $\left(1,\ -1\tfrac{1}{2}\right)$
- Ⓒ $(0,\ -3)$
- Ⓓ $\left(-1\tfrac{1}{2},\ -4\tfrac{1}{2}\right)$

2 For which point(s) do the *x*- and *y*-coordinates have the same value?

- Ⓕ Point *S* only
- Ⓖ Points *S* and *W* only
- Ⓗ Points *S*, *W*, and *X* only
- Ⓙ Points *S*, *W*, *X*, and *V* only

3 Points *T*, *S*, and *U* represent 3 vertices of a parallelogram. Which best represents point *Y*, the fourth vertex of the parallelogram?

- Ⓐ $\left(2,\ -3\tfrac{1}{2}\right)$
- Ⓑ $\left(1,\ -3\tfrac{1}{2}\right)$
- Ⓒ $\left(\tfrac{1}{2},\ -3\tfrac{1}{2}\right)$
- Ⓓ $\left(0,\ -3\tfrac{1}{2}\right)$

4 Which ordered pair represents a point located inside the quadrilateral but outside the pentagon?

- Ⓕ (-0.5, -1.25)
- Ⓖ (-0.25, 0.75)
- Ⓗ (-0.75, -0.5)
- Ⓙ (-1.5, -0.75)

5 Which of the following represents a point in Quadrant III that is located on the perimeter of the pentagon?

- Ⓐ (-1.5, -0.25)
- Ⓑ (-0.75, 0.25)
- Ⓒ (0.75, -1.5)
- Ⓓ (-1.25, -1.25)

Unit 33 Assessment

Standard 6.11(A) – Readiness

Use the grid to answer questions 1–6.

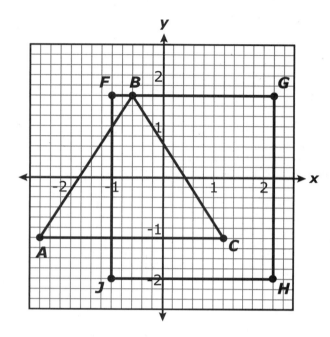

1 Which ordered pair represents a point inside both the triangle and the rectangle?

Ⓐ (0.2, 1)

Ⓑ (0.8, 0.4)

Ⓒ (-0.6, 1.2)

Ⓓ (-0.4, -1.4)

2 Which of the following represents a point in Quadrant IV that is located at a vertex of one of the figures?

Ⓕ (1.2, -1.2)

Ⓖ (-2.4, -1.2)

Ⓗ (2.2, -1.6)

Ⓙ (-0.6, 1.6)

3 Which best describes the signs of all coordinates located in Quadrant II?

Ⓐ (-x, -y)

Ⓑ (-x, +y)

Ⓒ (+x, -y)

Ⓓ (+x, +y)

4 A right triangle is formed using points C and H as two of the vertices. Which point best represents the coordinates for point X, the third vertex of the triangle?

Ⓕ (1.2, -1.8)

Ⓖ (1.8, -2.6)

Ⓗ (2.2, -1)

Ⓙ (2.2, -1.2)

5 Which ordered pair represents an intersection of two line segments?

Ⓐ (0, -1.6)

Ⓑ (-1.2, 1)

Ⓒ (-1, -1.2)

Ⓓ (0.8, -1.2)

6 Which ordered pair represents a point located inside the triangle but outside the rectangle?

Ⓕ (-0.2, 1.4)

Ⓖ (-1.2, -0.2)

Ⓗ (-0.4, -1.4)

Ⓙ (0.4, -0.8)

Name _____

Standard 6.11(A) – Readiness

Unit 33 Critical Thinking

Use the grid to answer the questions that follow.

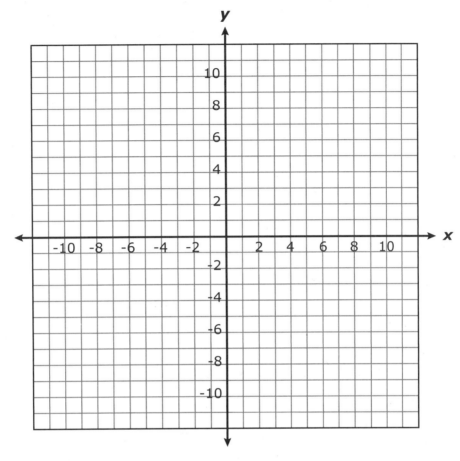

1. An ordered pair is located in Quadrant III. The x-coordinate is greater than the y-coordinate. List 3 possible ordered pairs that meet this criteria.

   _____   _____   _____

2. Draw a line segment on the coordinate plane above using the following criteria:
   - One endpoint must be located in Quadrant IV.
   - The line segment must intersect the y-axis but must NOT intersect the x-axis.

   In which quadrant or quadrants does the line segment lie? _____

   What are the endpoints of the line segment? _____

   If the x- and y-coordinates of the endpoint located in Quadrant IV are reversed, describe what happens to the point's location.

   _____

©2014 mentoringminds.com   motivationmath™ LEVEL 6   ILLEGAL TO COPY   267

Unit 33 Journal/Vocabulary Activity

Standard 6.11(A) – Readiness

Journal

Explain to a younger student what happens when an ordered pair is not plotted in the correct order.

_____
_____

Is there ever a time when the order of the coordinates does not matter? Explain.

_____
_____

Vocabulary Activity

Use the terms in the box to correctly label the picture shown. Each term is used only once.

x-coordinate	Quadrant III	coordinate plane	x-axis
y-axis	origin	Quadrant I	point	Quadrant IV
y-coordinate	Quadrant II	ordered pair

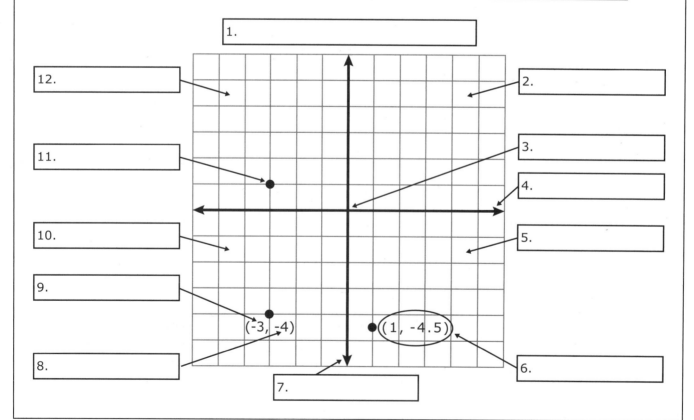

Get the Picture?

Complete *Get the Picture?* individually. Plot the ordered pairs listed below, and then connect them in the order they are shown to reveal a picture.

Ordered Pairs

1. $(3\frac{2}{3}, 2\frac{1}{3})$
2. $(3, 2)$
3. $(2\frac{1}{3}, \frac{1}{3})$
4. $(2, -1\frac{1}{3})$
5. $(2, -2\frac{2}{3})$
6. $(1\frac{1}{3}, -2\frac{2}{3})$
7. $(1\frac{1}{3}, -1\frac{1}{3})$
8. $(1, -1\frac{2}{3})$
9. $(\frac{2}{3}, -2\frac{2}{3})$
10. $(0, -2\frac{2}{3})$
11. $(\frac{1}{3}, -1\frac{2}{3})$
12. $(\frac{1}{3}, -1\frac{1}{3})$
13. $(-\frac{1}{3}, -1\frac{1}{3})$
14. $(-\frac{1}{3}, -2\frac{2}{3})$
15. $(-1, -2\frac{2}{3})$
16. $(-1, -2)$
17. $(-1\frac{1}{3}, -2\frac{2}{3})$
18. $(-2, -2\frac{2}{3})$
19. $(-1\frac{1}{3}, -1\frac{1}{3})$
20. $(-2, -1\frac{1}{3})$
21. $(-2\frac{2}{3}, -2)$
22. $(-3\frac{1}{3}, -2\frac{2}{3})$
23. $(-4, -2\frac{2}{3})$
24. $(-3\frac{2}{3}, -2\frac{1}{3})$
25. $(-3, -1\frac{2}{3})$
26. $(-2\frac{1}{3}, -\frac{1}{3})$
27. $(-1, \frac{1}{3})$
28. $(1\frac{1}{3}, \frac{1}{3})$
29. $(2\frac{2}{3}, 2\frac{1}{3})$
30. $(3\frac{1}{3}, 2\frac{2}{3})$
31. $(3\frac{2}{3}, 2\frac{2}{3})$

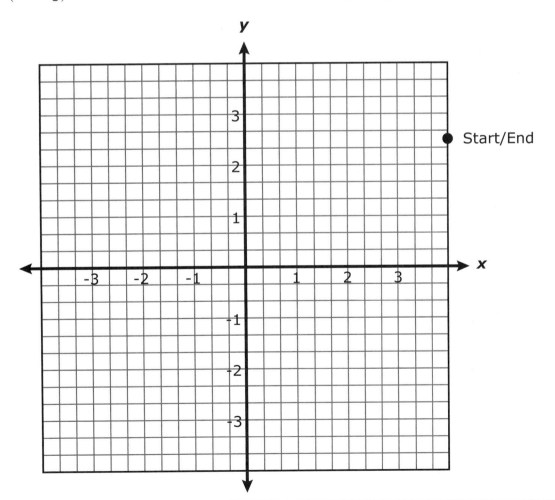

Unit 33 Homework

Standard 6.11(A) – Readiness

Use the coordinate grid to answer questions 1–9.

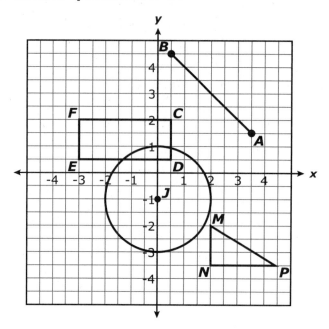

1. Plot a point that lies on $\overline{AB}$, and label it Q.

2. What are the coordinates for point Q?

3. List the ordered pairs for each labeled point that lies in Quadrant II.

4. List one ordered pair that lies on the circle and inside the rectangle.

5. List one ordered pair that lies inside the circle and that is located in Quadrant IV.

6. Plot a point that could be used to complete a rectangle that is twice the area of triangle MNP. Label the point R. What are the coordinates for point R?

7. Plot point (x, y) where $x < 0$ and $x \cdot y > 0$. Label the point S. Explain how you determined where to plot point S.

8. In which quadrant does point S lie?

9. In which quadrant does point J lie? Explain.

Connections

1. Research jobs that use the coordinate plane. Select 2 different jobs and write one paragraph about each, explaining how the coordinate plane is used and why it is important to that job. Share with the class.

2. Use string and stakes to create a coordinate plane in the yard. Take turns with friends and family tossing a beanbag, or a similar object, onto the grid. Give the coordinates of the location where the object lands. If correct, the person earns a point. The winner is the person with the most points.

Standard 6.12(A) – Supporting

Unit 34 Introduction

1. Mrs. Choy used a graphing calculator to generate 20 random numbers from 1 to 10. The results are displayed below.

2	3	8	5	4
8	10	8	2	4
6	6	4	3	3
9	2	6	4	8

Answer these questions using the data provided.

a. What is the minimum value? _____

b. What is the maximum value? _____

c. What is the median? _____

d. What is the value of the lower quartile (median of the first half)? _____

e. What is the value of the upper quartile (median of the second half)? _____

Create a box plot to display the data.

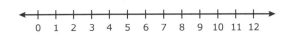

Use the chart to answer questions 2–4.

Anton records the different prices for bubble gum using the frequency chart below.

Gum Prices

$0.60	I	$1.10	II
$0.80	II	$1.20	III
$0.90	I	$1.30	II
$1.00	III	$1.50	II

2. Create a histogram to display the data.

3. Create a dot plot to display the data.

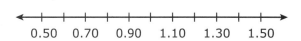

4. Create a stem-and-leaf plot using the data.

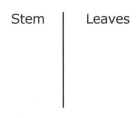

0 | 6 represents 0.6

Unit 34 Guided Practice

Standard 6.12(A) – Supporting

1 Hayden measured plant heights in centimeters and recorded the data as shown below.

18, 29, 17, 23, 21, 15, 24, 19, 27, 22, 20

He used the data to create a box plot. Which of the following box plots best represents Hayden's data?

Ⓐ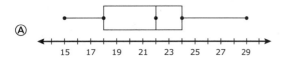

Ⓑ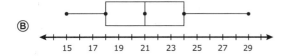

Ⓒ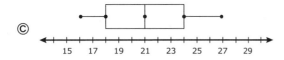

Ⓓ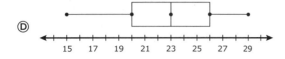

2 The Golden Retriever Kennel Club recorded the weight of each dog as shown below.

70, 57, 68, 80, 72, 72, 59, 67,

72, 65, 79, 68, 76, 63, 74

A stem-and-leaf plot is made using the recorded weights. Which does NOT correctly represent a line of stem and leaves from the plot?

Ⓕ 5 | 7 9

Ⓖ 6 | 3 5 7 8 8

Ⓗ 7 | 0 2 2 2 4

Ⓙ 8 | 0

3 Ms. Carlisle evaluated the semester project grades for her first-period class. She created the histogram shown below based on a dot plot of the project grades.

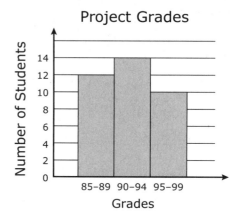

Which of the following dot plots could have been used to create the histogram?

Ⓐ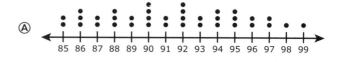

Ⓑ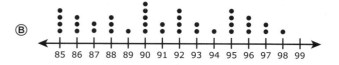

Ⓒ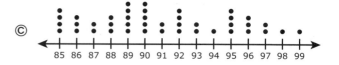

Ⓓ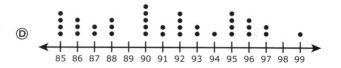

Name _____

Standard 6.12(A) – Supporting

Unit 34 Independent Practice

Use the information below to answer questions 1 and 2.

Brookside Grocery sponsored a team in the local charity run. The ages of the runners are shown below.

35, 23, 45, 21, 18, 19,

54, 46, 37, 25, 29, 18, 33

1   Which box plot best represents the ages of the runners?

Ⓐ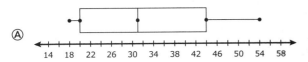

Ⓑ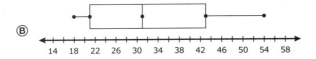

Ⓒ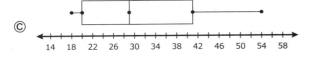

Ⓓ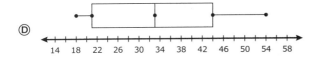

2   A histogram is made using the runners' ages. Which quantity best represents the bar on a histogram that shows the age range 25–34?

Ⓕ 2

Ⓖ 3

Ⓗ 4

Ⓙ 5

3   Remy made the following dot plot based on the shoe sizes of the girls in her class.

Shoe Sizes

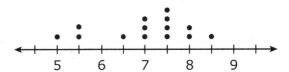

Which stem-and-leaf plot matches the data shown in Remy's dot plot?

Ⓐ
Stem	Leaves
5	0
5	5 5
6	5
7	0 0 0
7	5 5 5 5
8	0 0
8	5

5 | 0 represents 5.0

Ⓑ
Stem	Leaves
5	0 5 5
6	5
7	5
7	0 0 0 5 5 5
8	0 5 5

5 | 0 represents 5.0

Ⓒ
Stem	Leaves
5	0 5 5
6	5
7	0 0 0 5 5 5 5
8	0 0 5

5 | 0 represents 5.0

Ⓓ
Stem	Leaves
5	5 5
6	5
7	5 5 5 5
8	5

5 | 0 represents 5.0

Unit 34 Assessment

Standard 6.12(A) – Supporting

1 Ginger records her grades for each assignment in science.

88	100	65	84	95
92	78	98	89	78
57	70	89	86	

Which graphical display correctly represents Ginger's science grades?

Ⓐ
Stem	Leaves
5	7
6	5
7	0 8 8
8	4 6 8 9 9
9	2 5 8
1	0 0

5 | 7 represents 57

Ⓑ

Ⓒ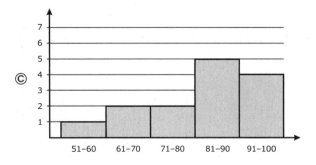

Ⓓ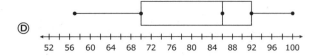

2 Brandon has a job mowing lawns. His earnings are represented in the box plot shown below.

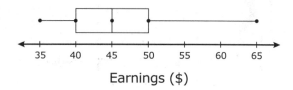

Earnings ($)

Which chart best corresponds to the information provided in the box plot?

Ⓕ
Dollar Amount	# of Jobs
$35	1
$40	4
$45	0
$50	2
$55	1
$60	1
$65	2

Ⓗ
Dollar Amount	# of Jobs
$35	2
$40	2
$45	2
$50	1
$55	2
$60	1
$65	1

Ⓖ
Dollar Amount	# of Jobs
$35	3
$40	2
$45	1
$50	2
$55	1
$60	1
$65	1

Ⓙ
Dollar Amount	# of Jobs
$35	2
$40	3
$45	2
$50	2
$55	1
$60	0
$65	1

Name _____

Standard 6.12(A) – Supporting

Unit 34 Critical Thinking

Shown below are two graphs representing the same data set.

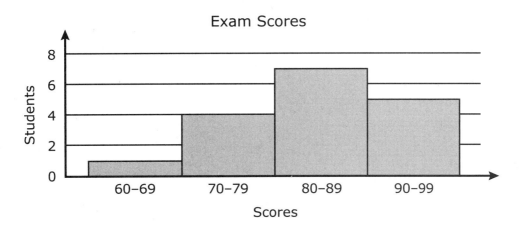

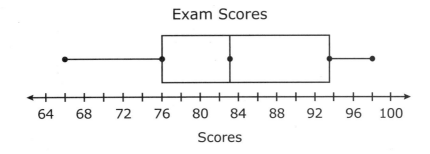

Answer the following questions about the graphs.

1. What are the advantages of using the histogram instead of the box plot to represent the data?
   _____
   _____

2. What are the advantages of using the box plot instead of the histogram to represent the data?
   _____
   _____

3. What are the disadvantages, if any, of using both the histogram and the box plot to represent the data?
   _____
   _____

Unit 34 Journal/Vocabulary Activity

Name _____

Standard 6.12(A) – Supporting

Journal

Explain why there are no gaps between the bars of a histogram.

_____
_____
_____

Vocabulary Activity

For each example shown, record the type of graph in the blank above the example and label the parts of the graph as indicated.

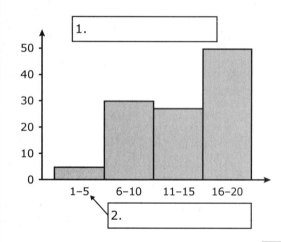

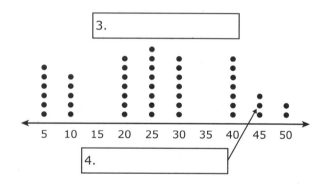

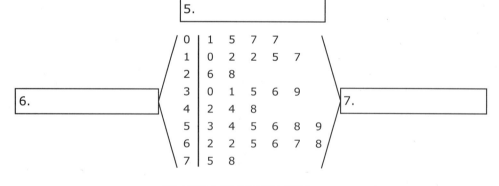

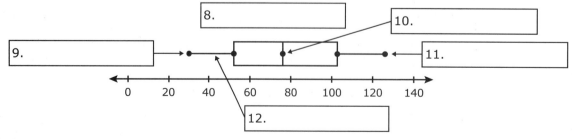

Standard 6.12(A) — Supporting

Unit 34 Motivation Station

Name that Graph

Play *Name that Graph* with a partner. Each pair of players needs a game board and 4 sticky notes numbered 1–4. The sticky notes are placed over the clue boxes below in random order. Players should not read the clues before the game begins. Player 1 begins by stating the number of clues (1, 2, or 3) he/she believes will be needed to identify the type of graph being described by player 2. Player 2 begins with the clue box covered with sticky note 1, reading one clue at a time until player 1 guesses the type of graph being described. If the graph is identified correctly in the number of clues stated or less, the number of points earned is as follows:

- One clue used to identify – 30 points
- Two clues used to identify – 25 points
- Three clues used to identify – 15 points

If the graph is incorrectly identified, or not identified in the number of clues stated, the player loses 10 points. Players switch roles and repeat the process for the remaining clue boxes. The winner is the player with more points after all four boxes are revealed.

Clue 1: This graph makes it easy to see the trends in a data set.

Clue 2: This graph does not show individual data points.

Clue 3: This graph uses bars with no spaces between to display continuous data.

Answer: histogram

Clue 1: This graph makes the median easy to identify.

Clue 2: This graph is graphed on a number line.

Clue 3: This graph contains 50% of the data in a box.

Answer: box plot

Clue 1: This graph makes it easy to see the trends in a data set.

Clue 2: This graph shows individual data points.

Clue 3: This graph uses place value to graph data.

Answer: stem-and-leaf plot

Clue 1: This graph shows individual data points.

Clue 2: This graph is graphed on a number line.

Clue 3: This graph uses individual dots to represent data points.

Answer: dot plot

Unit 34 Homework

Standard 6.12(A) – Supporting

1. Is it possible for an interval on a histogram to have a frequency of 0? If so, explain what to look for in the graph. If not, explain your reasoning.

   _____
   _____
   _____

Use the box plot to answer questions 2–5.

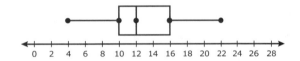

2. What is the minimum value in the plot?

   _____

3. What is the maximum value in the plot?

   _____

4. What is the median?

   _____

5. What are the values of each quartile?

   1st _____

   2nd _____

   3rd _____

6. The following data set shows the number of points the Trojans scored in each basketball game this season.

   89, 57, 42, 59, 75, 70, 39, 54, 57,

   86, 33, 79, 52, 55, 67, 60, 59, 52

   Represent the Trojans' scores using a stem-and-leaf plot and then a histogram. Label both graphs.

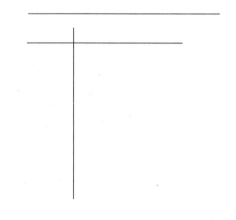

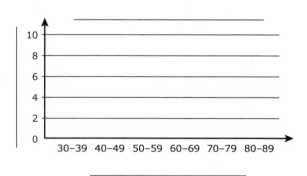

Connections

1. Visit the website http://illuminations.nctm.org to find interactive, online graphing tools. Use the search feature to find the Advanced Data Grapher and the Histogram Tool. Input your own data and explore creating and learning more about different types of graphs.

2. Search newspapers and online resources to find an example of each type of graph studied in this unit. Attach the graphs to a sheet of paper, and explain the data found in each type. Share your findings with the class.

Name _____

Standard 6.12(B) – Supporting

Unit 35 Introduction

1. The lunch account balances of Mrs. Miller's students are shown in the dot plot below.

Lunch Account Balances

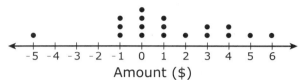

Amount ($)

Circle the word that best describes the shape of the distribution.

   Skewed   Symmetric   Uniform

Describe the peaks, gaps, outliers, and spread shown in the data distribution.

Peak: _____

Gap: _____

Outlier: _____

Spread: _____

Which measure of center is most appropriate to describe the data distribution? Circle your answer.

   Mean   Median

Explain.

_____

_____

2. Look at the histogram below.

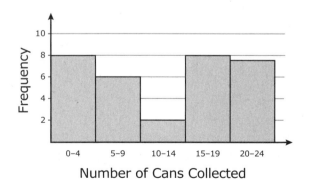

Number of Cans Collected

For which intervals are there peaks in the data distribution?

_____

3. The box plot below shows the number of correct answers on a 20-question quiz for 15 students.

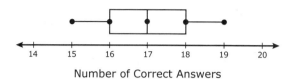

Number of Correct Answers

Describe the shape of the distribution.

_____

Does the data appear to have an outlier? Explain your reasoning.

_____

_____

One measure of center for a data distribution is the median. What is the median of this data? _____

What is the mean of this data? _____

Explain how you know.

_____

_____

4. Create a stem-and-leaf plot using the information below.

   • The data has a minimum value of 15.
   • The data shows a spread of 24.
   • The plot is uniform in shape.
   • The plot contains exactly 15 data points.

Stem	Leaves
1	
2	
3	

1 | 5   represents 15

Unit 35 Guided Practice

Standard 6.12(B) – Supporting

Use the graph to answer questions 1 and 2.

North Beach Middle School participated in a fund-raiser to raise money for new athletic equipment. The results of the fund-raiser are displayed below.

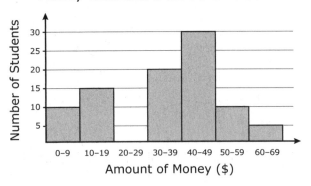

1. Which best describes the spread of the data?

    Ⓐ $40–$49

    Ⓑ $20–$49

    Ⓒ $30–$69

    Ⓓ $0–$69

2. Which statement is true about the center of the distribution?

    Ⓕ The center is best described using the mean, because the distribution is skewed.

    Ⓖ The center is best described using the median, because the distribution is skewed.

    Ⓗ The center is best described by the mean and median, because the distribution is symmetric.

    Ⓙ The center is best described by the outlier, because the distribution is symmetric.

Use the graph to answer questions 3 and 4.

Darrien researched the ages of some of his favorite professional football players. He recorded his results in the dot plot below.

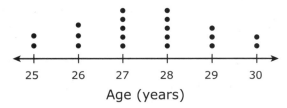

3. Which of the following statements about the distribution are true?

    I. There are two peaks at 27 and 28.

    II. The distribution is skewed.

    III. The distribution does not have an outlier.

    IV. The distribution shows a spread of 5.

    Ⓐ I only

    Ⓑ I, II, and III only

    Ⓒ I, III, and IV only

    Ⓓ I, II, III, and IV

4. Based on the shape of the data distribution, which statement is true about the measure of the center?

    Ⓕ The mean and median are equivalent.

    Ⓖ The mean is greater than the median.

    Ⓗ The mean is less than the median.

    Ⓙ Not here

Name _____

Standard 6.12(B) – Supporting

Unit 35 Independent Practice

Use the graph to answer questions 1 and 2.

Ramona conducts a survey of sixth-grade students to determine the number of pages they can read in one hour. The results are shown in the stem-and-leaf plot.

Stem	Leaves
2	2 7 8 8 9
3	1 3 5 5 6 8 9 9 9
4	2 8 9 9
5	0 1
7	0

2 | 2   represents 22

1 Which of the following statements does NOT support the data shown in the stem-and-leaf plot?

Ⓐ The shape of the distribution is skewed.

Ⓑ The spread of the distribution is 48.

Ⓒ There are no outliers represented in the distribution.

Ⓓ There is a single peak in the distribution.

2 Which of the following is true?

Ⓕ The mean and median are equivalent.

Ⓖ The median best describes the center of the distribution.

Ⓗ The distribution of the data is uniform.

Ⓙ All of the above

Use the graph to answer questions 3–5.

Mr. Nichols uses a box plot to show the number of e-mails he receives each day during one week.

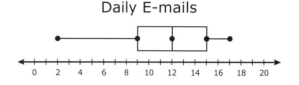

Daily E-mails

3 Could the data contain an outlier?

Ⓐ Yes, because the distribution of data is symmetric

Ⓑ No, because the distribution of data is skewed to the right

Ⓒ Yes, because the distribution of data is not symmetric

Ⓓ There is not enough information to determine.

4 Which measure best describes the center of the data distribution?

Ⓕ Spread

Ⓖ Mean

Ⓗ Peak

Ⓙ Median

5 Which value best describes the spread?

Ⓐ 15

Ⓑ 6

Ⓒ 20

Ⓓ 12

Unit 35 Assessment

Standard 6.12(B) – Supporting

Use the graph to answer questions 1 and 2.

Raymond's club soccer team sells raffle tickets for a flat screen television. The dot plot shows the number of tickets each player sold.

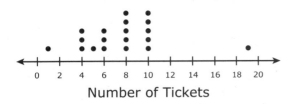

Raffle Tickets Sold
Number of Tickets

1. Which of the following statements are true?

   I. The distribution is skewed.

   II. The value 19 is an outlier.

   III. There are two peaks at 8 and 10.

   IV. There is a gap from 11 to 18.

   Ⓐ IV only

   Ⓑ II and III only

   Ⓒ I, II, and IV only

   Ⓓ I, II, III, and IV

2. Which statement is true about the center of the data distribution?

   Ⓕ The center is best described by the peak.

   Ⓖ The center is best described by the median.

   Ⓗ The center is best described by the outlier.

   Ⓙ The center is best described by the mean.

3. Look at the histogram below.

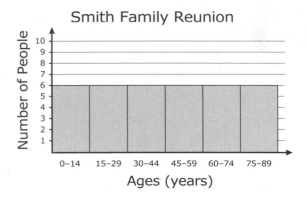

Smith Family Reunion
Ages (years)

Which is true about the shape and spread of the data distribution?

Ⓐ Uniform, with a spread of 89 years

Ⓑ Skewed, with a spread of 45 years

Ⓒ Symmetric, with a spread of 14 years

Ⓓ Skewed, with a spread of 89 years

4. Elijah kept a journal of the amount of rainfall he received at his house each day during the month of April. The data is displayed in the box plot below.

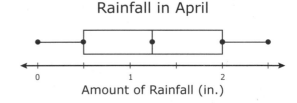

Rainfall in April
Amount of Rainfall (in.)

Which of the following is NOT true?

Ⓕ The mean and median are equivalent.

Ⓖ The data contains no outliers.

Ⓗ The value of the center is 1.5.

Ⓙ The spread of the data distribution is from 0 to 2.5.

Name _____

Standard 6.12(B) – Supporting

Unit 35 Critical Thinking

Create a graph to fit each set of criteria.

1. Stem-and-leaf plot with uniform shape, a spread of 20, and containing 2 data points

2. Dot plot with most data on the left side (skewed right), with a spread of 10, one outlier, and containing 15 data points

3. Box plot with symmetric shape, a median of 12, and a spread of 16

4. Histogram with most data on the right side (skewed left), with a total frequency of 15, and a spread of 1 to 20

Unit 35 Journal/Vocabulary Activity

Name _____

Standard 6.12(B) – Supporting

Journal

Explain the three ways used to describe the distribution of data in a set.

_____
_____
_____
_____

Vocabulary Activity

For each term, sketch a visual representation and explain the meaning of the term. Use any graph to create the representation: histogram, dot plot, stem-and-leaf plot, or box plot.

1. Uniform distribution

2. Symmetric distribution

3. Skewed distribution

4. Spread of distribution

5. Outlier(s)

Name _____

Standard 6.12(B) – Supporting

Unit 35 Motivation Station

Descriptive Decisions

Play *Descriptive Decisions* with a partner. Each pair of players needs a game board. Each player needs a pencil. Player 1 chooses any problem to complete, matching the description to the correct graph. If correct, player 1 writes the problem number and his/her initials in the box. If incorrect, player 1 loses a turn. Play passes to player 2 who repeats the process. Play continues until one player gets 3 boxes in a row, either horizontally, vertically, or diagonally, or until all boxes are initialed.

1. Skewed right, spread = 10
2. Uniform, spread = 20
3. Skewed left, spread = 23
4. Symmetric, spread = 10
5. Skewed right, spread = 25
6. Uniform, spread = 16
7. Skewed left, spread = 9
8. Symmetric, spread = 16

9. Skewed right, spread = 16
10. Uniform, spread = 14
11. Skewed left, spread = 16
12. Symmetric, spread = 20
13. Skewed right, spread = 27
14. Symmetric, spread = 24
15. Skewed left, spread = 15

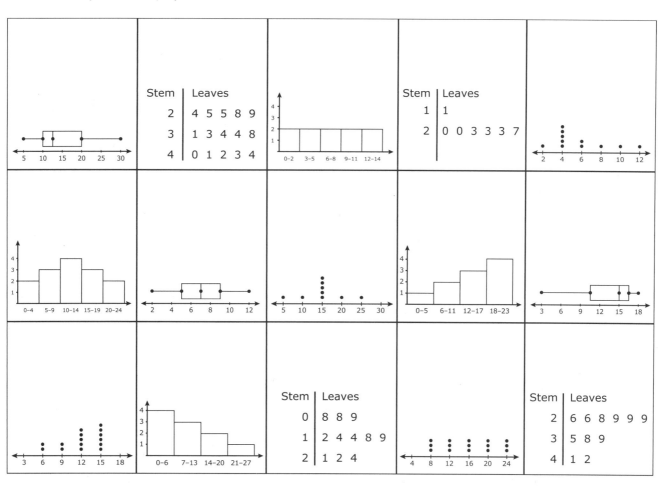

Unit 35 Homework

Standard 6.12(B) – Supporting

1 Create a box plot using the following criteria:
- 25% of the data is between 12 and 18
- 50% of the data is between 18 and 24
- 25% of the data is between 24 and 28
- The median is 22.

Describe the shape of the data's distribution.
_____

What is the spread of the data's distribution?
_____

2 A stem-and-leaf plot containing 20 data points is uniform in shape and has a spread of 25. The minimum value is between 5 and 9. Create a possible plot below.

Stem	Leaves

Which measure of central tendency best describes the center of the data distribution? Explain.
_____
_____
_____

3 The dot plot below shows the number of children in each student's family.

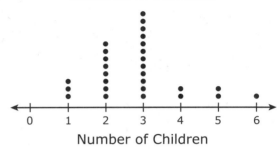

Which measure of central tendency best describes the center of the data distribution?
_____

Describe the shape of the data's distribution.
_____

4 Look at the histogram below.

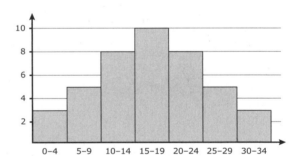

Describe the shape of the data's distribution.
_____

What is the spread of the data's distribution?
_____

Connections

Using the Internet, research what a bell curve means in statistics. How is the bell curve related to the shape of data you have studied in this unit? Summarize your findings to share with the class.

Standard 6.12(C) – Readiness

Unit 36 Introduction

Use the information below to answer questions 1–6.

Hector counted the number of candy pieces in 15 fun-size bags. His results are recorded below.

12	14	14	13	13
11	12	13	15	14
13	12	12	13	14

1. What is the median of the data? _____

2. What is the mean of the data? _____

3. Write a statement that compares the values of the measures of center.

4. What conclusion can be made about the shape of the data distribution based on the measures of center? Justify your answer.

5. Determine the spread of the data distribution using the range.

   _____

6. Determine the spread of the data distribution using the interquartile range (IQR).

   _____

Use the information below to answer questions 7–9.

Rodney has earned the following grades:

Math: 80, 90, 75, 80, 100

Science: 75, 80, 65, 85, 90, 85

English: 100, 95, 100, 55, 90, 100

7. The data for which subject best matches the measures of center and spread shown below?

   Mean = 80

   Median = 82.5

   Range = 25

   IQR = 10

   _____

8. Calculate the values for the measures of spread for Rodney's English grades.

   Range _____   IQR _____

   Which measure is best used to describe the spread of the data for Rodney's English grades? Explain.

9. What range of values describes the middle 50% of Rodney's grades in math?

   _____

Unit 36 Guided Practice

Standard 6.12(C) – Readiness

Use the information below to answer questions 1 and 2.

The Cumberland Knights were one of eight basketball teams that competed in the quarterfinals. Each team's final score is represented below.

50, 70, 71, 72, 73, 72, 62, 72

1 Which of the following is true about the final scores?

 I. The spread of the scores can be described by the value 23.

 II. A median of 72.5 can be used to describe the center.

 III. The shape of the data is skewed because the mean and median are not equivalent.

 IV. The interquartile range has a value of 6.

 Ⓐ I only

 Ⓑ II and III only

 Ⓒ III and IV only

 Ⓓ I, III, and IV only

2 What range of values describes the middle 50% of the scores?

 Ⓕ 71 to 72

 Ⓖ 66 to 72

 Ⓗ 69 to 71

 Ⓙ 66 to 71.5

Use the information below to answer questions 3 and 4.

Raul and Bryan each recorded the number of hours they slept last week. The data is shown below.

Raul's Sleep Log

8.5   8   9.5   9   8.5   7.5   8.5

Bryan's Sleep Log

6.5   7   7.5   6.5   12   7   7.5

3 Which statement is NOT true about the data provided?

 Ⓐ Both sets of data have an interquartile range of 1.

 Ⓑ Both measures of center for Raul's data have a value of 8.5.

 Ⓒ The shapes of both data distributions are non-symmetric.

 Ⓓ The IQR of Bryan's data is best used to describe the spread because of the outlier.

4 Which is a true comparison of the measures of center for Bryan's data?

 Ⓕ Range = Mean

 Ⓖ Mean < Median

 Ⓗ Median > IQR

 Ⓙ Mean > Median

Name _____

Standard 6.12(C) – Readiness

Unit 36 Independent Practice

Use the information below to answer questions 1-3.

Consuelo recorded the amount of time, in minutes, she spent exercising on the treadmill.

15, 25, 30, 25, 45, 20, 35, 30, 35, 30, 40

1 Which value best represents the interquartile range?

Ⓐ 30 min

Ⓑ 25 min

Ⓒ 10 min

Ⓓ 35 min

2 Which shows the correct values for the measures of center?

Ⓕ Mean = 30; Median = 30

Ⓖ Range = 30; Median = 20

Ⓗ Median = 30; Mean = 20

Ⓙ Mean = 25; Range = 30

3 What conclusion can be drawn about the shape of the data distribution based on the measures of center?

Ⓐ The shape of the distribution is best described as uniform.

Ⓑ The shape of the distribution is best described as symmetric.

Ⓒ The shape of the distribution is best described as skewed.

Ⓓ Not enough information is given to determine the shape of the distribution.

Use the information below to answer questions 4 and 5.

Mr. Brooks held auditions for the school musical each afternoon. The data below shows the number of students who auditioned each day.

22, 18, 20, 26, 24, 2, 18, 21

4 Which correctly describes the spread of the data?

Ⓕ Range = 24
IQR = 20.5

Ⓖ Range = 21
IQR = 5

Ⓗ Range = 24
IQR = 5

Ⓙ Range = 20.5
IQR = 21

5 Based on the data provided, which is best used to describe the spread?

Ⓐ Range, because the data is skewed in shape

Ⓑ IQR, because there are an even number of data points

Ⓒ Range, because the minimum value is 2 and the maximum value is 26

Ⓓ IQR, because the data contains an outlier

Unit 36 Assessment

Name _____

Standard 6.12(C) – Readiness

1 In which of the following data sets are the measures of center equivalent?

  Ⓐ {6, 4, 3, 2, 1, 4, 2, 3}

  Ⓑ {3, 1, 2, 6, 4, 3, 2, 3}

  Ⓒ {4, 3, 3, 2, 2, 1, 2, 1}

  Ⓓ {2, 4, 6, 6, 3, 1, 2, 2}

Use the information below to answer questions 2 and 3.

Mr. Miller shops for a new printer. He searches the local ads for printers and records the prices as shown below.

$150, $129, $79, $145, $136, $140, $130

2 What value best describes the median?

  Ⓕ $129

  Ⓖ $145

  Ⓗ $136

  Ⓙ $140

3 Which value best describes the measure of spread for the prices?

  Ⓐ $70, because that is the value that represents the range

  Ⓑ $136, because that is the middle value of the data

  Ⓒ $89, because that is the difference between the minimum and maximum values

  Ⓓ $16, because the data shows an outlier

Use the information below to answer questions 4-6.

The data shows the time in seconds for 8 runners competing in the 100-meter dash at Saturday's track meet.

14, 13, 16, 15, 12, 14, 13, 15

4 What value describes the mean of the data?

  Ⓕ 13

  Ⓖ 13.5

  Ⓗ 14

  Ⓙ 14.5

5 Which statement is true about the measures of center?

  Ⓐ The median and mean are equivalent, so the shape of the distribution is symmetric.

  Ⓑ The mean is greater than the median, so the shape of the distribution is skewed.

  Ⓒ The median is greater than the mean, so the shape of the distribution is non-symmetric.

  Ⓓ You cannot determine any information about the shape of the distribution using the measures of center.

6 Which of the following could NOT be used to describe a measure of spread?

  Ⓕ 4

  Ⓖ 13 to 15

  Ⓗ 12 to 16

  Ⓙ 14

Name _____

Standard 6.12(C) – Readiness

Unit 36 Critical Thinking

Mr. Alexander gives a benchmark assessment to all of his classes. The assessment results for two of the classes are shown below.

Class A: 84, 100, 75, 68, 92, 94, 60, 100, 70, 86, 94, 96, 91, 90

Class B: 84, 70, 71, 60, 91, 73, 77, 92, 86, 60, 75, 76, 85, 73

1 What are the measures of center for each class? Round answers to the nearest tenth.

Class A: Mean _____ Median _____

Class B: Mean _____ Median _____

Based on the values for the measures above, what conclusion(s) can be drawn about the shape of each data distribution? Create a double dot plot to justify your response.

_____

_____

Class A

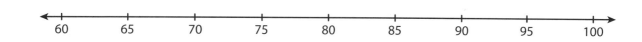

Class B

2 What are the values for the measures of spread for each class?

Class A: Range _____ Interquartile Range (IQR) _____

Class B: Range _____ Interquartile Range (IQR) _____

What do the IQR values for each class reveal about the data distribution around the median?

_____

_____

Unit 36 Journal/Vocabulary Activity

Name _____

Standard 6.12(C) – Readiness

Journal

When is the interquartile range a better indicator of the spread of data distribution than the range? Explain.

_____
_____
_____
_____

Vocabulary Activity

Complete each row of the chart for the vocabulary term given.

Term	Non-math Definition	Math Definition	Example/Representation
Mean			
Median			
Range			
Outlier			

292 ILLEGAL TO COPY motivation**math**™ LEVEL 6 ©2014 mentoring**minds**.com

Name _____

Standard 6.12(C) – Readiness

Unit 36 Motivation Station

Center on Color: Mean and Median

For each section in the picture below, a data set is given. Color the sections according to the rules given at the bottom of the page. Use a calculator as needed.

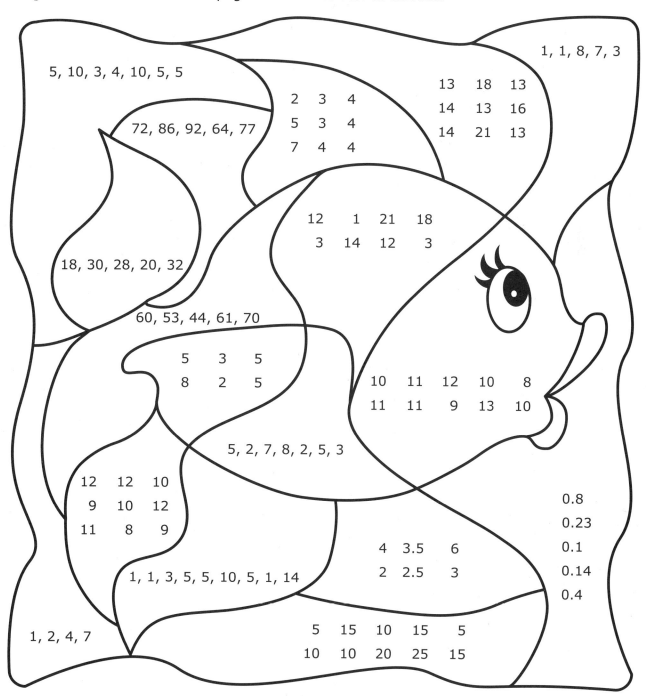

If the measures of center are equivalent, color the section yellow.

If the mean is greater than the median, color the section blue.

If the median is greater than the mean, color the section red.

Unit 36 Homework

Standard 6.12(C) – Readiness

Use the information to answer questions 1–3.

The heights of various cactus plants are shown below.

Cactus Heights (ft)

2	6	6	10	15	2	15
6	20	30	1	6	6	3

1. Calculate the values for the measures of center. Round answers to the nearest tenth.

   Mean _____   Median _____

   Complete the following sentence about the measures of center and the shape of the data distribution.

   The _____ is greater than the _____, so the shape of the data distribution is _____.

2. Express the measures of spread for the data set. Write each measure two different ways.

   _____

   _____

3. Based on the values for the measures of spread, what conclusion(s) can be made about the distribution of the data around the center?

   _____

   _____

4. The students in Ms. Heffler's first-period class earned the following scores on the most recent quiz.

   Quiz Scores

77	80	63	85
79	61	89	74
83	71	74	69
94	90	81	44

   While grading quizzes for another class, Ms. Heffler discovered an answer she marked as incorrect on the quiz that was actually correct. The mistake resulted in everyone receiving 4 additional points on their quiz grades. What effect does adding the additional 4 points to each student's score have on each measure?

Measure	Original Scores	New Scores	Effect
Mean			
Median			
Range			
IQR			

5. Based on your responses in the table above, what effect does the change in quiz scores have on the shape of the data distribution?

   _____

   _____

Connections

Use the Internet to research and find the average high temperatures for the month of January in your area over the last 20 years. Calculate the measures of center and the measures of spread for the data collected. What conclusions can be drawn from the values calculated? What do the values predict for the next 20 years? Summarize your findings to share with the class.

Name _____

Standard 6.12(D) – Readiness

Unit 37 Introduction

Use the information below to answer questions 1–4.

A class of freshman students at Lincoln High School choose their preferred university to attend after graduation. The results of the survey are shown below.

University Survey Results

University	Number of Students
University of Texas	IIII
Texas A&M	IIII I
Baylor	IIII
Texas Tech	III
Texas State	I
Rice	II

1 Use the information in the chart to complete the relative frequency table. Represent the relative frequency as a decimal to the nearest hundredth.

University Survey Results

University	Frequency	Relative Frequency
University of Texas		
Texas A&M		
Baylor		
Texas Tech		
Texas State		
Rice		

2 Which university represents the mode for the data shown?

_____

3 Write *true* or *false* for each statement based on the data from the frequency table.

Fifty percent of the students choose either Baylor or Texas A&M.

_____

One-fourth of the students choose to attend the University of Texas.

_____

Fewer than 80% of the students do not choose to attend Texas Tech.

_____

The three universities that were chosen by the fewest number of students make up $\frac{3}{10}$ of the results.

_____

4 Using the information from the frequency table, make a percent bar graph. Use six different colors to represent each of the universities.

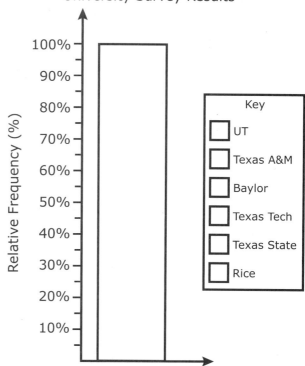

©2014 mentoringminds.com   motivationmath™ LEVEL 6   ILLEGAL TO COPY   295

Unit 37 Guided Practice

Standard 6.12(D) – Readiness

Use the information below to answer questions 1 and 2.

Mrs. Reynolds plants several different colored roses. If all of the rose plants bloom, she will have 2 yellow, 7 red, 5 pink, 7 white, and 4 orange rose plants.

1. Which color of rose represents the mode?

    Ⓐ Red only

    Ⓑ Red and pink only

    Ⓒ White only

    Ⓓ White and red only

2. Which values best complete the relative frequency table?

Color	Frequency	Relative Frequency
Yellow	2	
Red	7	0.28
Pink	5	0.20
White	7	0.28
Orange	4	

   Ⓕ Yellow = 0.20; Orange = 0.40

   Ⓖ Yellow = 0.08; Orange = 0.16

   Ⓗ Yellow = 0.06; Orange = 0.12

   Ⓙ Yellow = 0.80; Orange = 0.16

Use the graph to answer questions 3 and 4.

The farmers' market conducts a survey to determine the favorite local produce. The results are shown in the graph below.

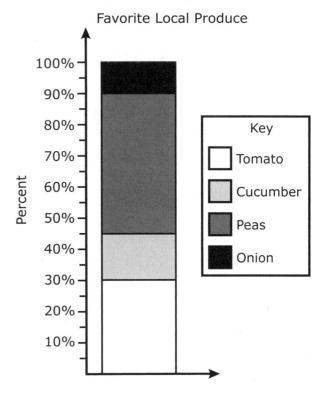

3. Which statement is true about the distribution of the data?

   Ⓐ Tomatoes and onions were selected as the favorite produce by over half of the people surveyed.

   Ⓑ Tomatoes represent the mode for the data.

   Ⓒ Of the people surveyed, $\frac{9}{20}$ chose peas as their favorite.

   Ⓓ Cucumbers were selected as the favorite produce by 45% of the people surveyed.

4. If 60 people were surveyed, which produce does NOT reflect the correct frequency?

   Ⓕ Tomato – 20   Ⓗ Peas – 27

   Ⓖ Cucumber – 9   Ⓙ Onion – 6

Standard 6.12(D) — Readiness

Unit 37 Independent Practice

Use the information below to answer questions 1–3.

The Jackson family has a large assortment of pets on their farm. The table below shows the number of each type of pet the family owns.

Pet	Bird	Cat	Dog	Snake
Frequency	3	5	11	1

1. Which information best completes a relative frequency table about the pets at the Jackson family farm?

   Ⓐ
Pet	Bird	Cat	Dog	Snake
Relative Frequency	0.20	0.25	0.45	0.10

   Ⓑ
Pet	Bird	Cat	Dog	Snake
Relative Frequency	0.15	0.20	0.55	0.10

   Ⓒ
Pet	Bird	Cat	Dog	Snake
Relative Frequency	0.15	0.25	0.55	0.05

   Ⓓ
Pet	Bird	Cat	Dog	Snake
Relative Frequency	0.15	0.20	0.60	0.05

2. Which of the following statements does NOT reflect the distribution of the data?

   Ⓕ One-fourth of the pets on the farm are cats.

   Ⓖ Snakes represent 10% of the pets on the farm.

   Ⓗ The number of birds and snakes on the farm make up 20% of the pets.

   Ⓙ More than half of the pets on the farm are dogs.

3. Which bar correctly shows the distribution of pets?

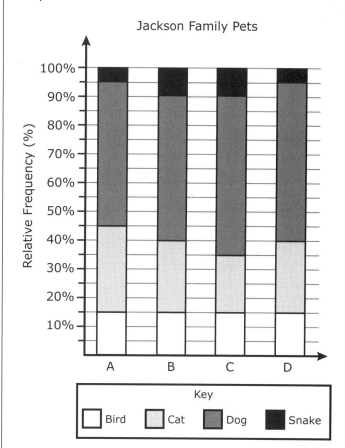

   Ⓐ Bar A

   Ⓑ Bar B

   Ⓒ Bar C

   Ⓓ Bar D

Unit 37 Assessment

Standard 6.12(D) – Readiness

Use the information below to answer questions 1–3.

Ridgeway Middle School offers after-school recreational sports. The graph below represents all the students in Mr. Bryant's first-period class and which sport they signed up to play after school.

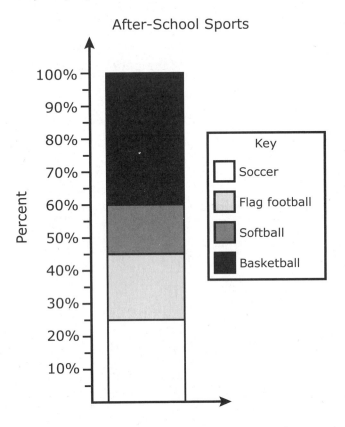

1. Which of the following statements best describes the distribution of data?

   Ⓐ Forty-five percent of the students in Mr. Bryant's class signed up to play softball.

   Ⓑ Less than one-third of the students in Mr. Bryant's class signed up to play soccer.

   Ⓒ More than $\frac{3}{5}$ of the students in Mr. Bryant's class signed up to play either flag football or soccer.

   Ⓓ Exactly $\frac{5}{8}$ of the students in Mr. Bryant's class did not sign up to play basketball.

2. Which sport represents the mode?

   Ⓕ Soccer Ⓗ Softball

   Ⓖ Flag football Ⓙ Basketball

3. Which relative frequency table best represents sports chosen by the students in Mr. Bryant's class?

Ⓐ
Sport	Relative Frequency
Soccer	0.25
Flag football	0.15
Softball	0.20
Basketball	0.40

Ⓑ
Sport	Relative Frequency
Soccer	0.25
Flag football	0.40
Softball	0.20
Basketball	0.15

Ⓒ
Sport	Relative Frequency
Soccer	0.25
Flag football	0.20
Softball	0.15
Basketball	0.40

Ⓓ
Sport	Relative Frequency
Soccer	0.25
Flag football	0.20
Softball	0.40
Basketball	0.15

Name _____

Standard 6.12(D) – Readiness

Unit 37 Critical Thinking

The cafeteria staff at Barker Middle School conducted a survey of sixth-grade students eating in the cafeteria to determine their likes and dislikes about the foods served. One of the survey questions and the data collected are shown below.

If the cafeteria had one day each week that was a particular food theme day, which of the following would you most like to see?

	Macaroni Monday	Taco Tuesday	Waffle Wednesday	Tater Tot Thursday	Fish Stick Friday
Sixth Grade	|||| |||| |||| |||| |||| |||	|||| |||| |||| |||| |||| |||| |||	|||| |||| |||| ||||		|||| |||| |||| |||| ||||		|||| |||| |||| ||

1 Use the data from the chart to complete the relative frequency table below. Round to the nearest hundredth.

Day	Macaroni Monday	Taco Tuesday	Waffle Wednesday	Tater Tot Thursday	Fish Stick Friday
Frequency					
Relative Frequency					

2 The same survey was given to seventh and eighth graders. The graph below shows the results of the survey for those grades. Complete the bar for sixth grade.

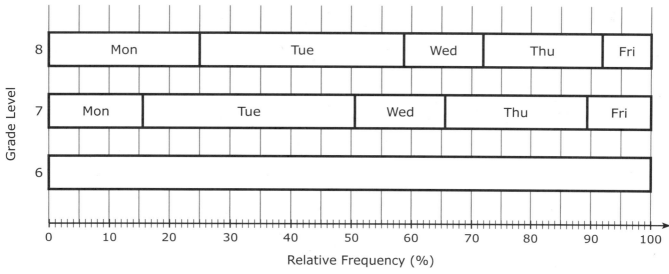

3 Write three statistical observations based on the data gathered from the survey.

_____

_____

_____

Unit 37 Journal/Vocabulary Activity

Standard 6.12(D) – Readiness

Journal

Explain how to find the mode from a relative frequency table and a percent bar graph.

_____
_____
_____

Vocabulary Activity

For each statement, identify the data described as a *frequency* or a *relative frequency*. Record your responses in the blanks.

1. Jeremy correctly answers 7 questions on the quiz. _____

2. Collette ate 2 hot dogs at the party. _____

3. Eight out of 10 doctors recommend brushing twice daily. _____

4. Roger lost 20 pounds of his 50-pound goal in six weeks. _____

5. Twelve students in Mr. Carney's class are out with the flu. _____

6. There are 3 defective widgets out of every 5 made in the factory. _____

7. Two of the three pens in Nell's purse are red. _____

8. Austin spent $10 of his birthday money on fast food. _____

9. Julie's family drove 350 miles of their trip on the first day. _____

10. Ten animals at the shelter were cats. _____

11. The temperature was above 100°F on 28 of the 31 days in July. _____

12. There are 14 boys in Mrs. Green's class of 23 students. _____

13. The Viking football team won 8 games last season. _____

14. Twelve out of 18 of Mr. Garcia's students made the honor roll. _____

15. For every 5 books in the school library, 3 are novels. _____

Categorical Summaries

Play *Categorical Summaries* with a partner. Each pair of players needs a game board and 2 paper clips to use with the spinners. Each player needs a pencil and 4 colored pencils (red, yellow, blue, green). Player 1 begins by spinning both spinners. The first spinner identifies the table, and the second spinner identifies the column to be completed. Player 1 records the relative frequency as a decimal in the table, then shades the bar on the graph to match the information and color in the table. If completed correctly, player 1 initials the box completed in the table and earns a number of points equal to the frequency of the category completed. If completed incorrectly, player 2 may steal the points if he/she can correct the error. Player 2 then continues with a turn. The game ends when all categories and the percent bar graph are complete. The player with more points wins.

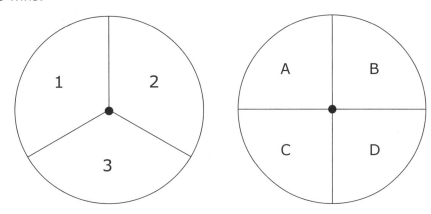

	A	B	C	D
1	Pepperoni (Red)	Cheese (Yellow)	Sausage (Blue)	Veggie (Green)
Frequency	10	9	3	3
Relative Freq.				

	A	B	C	D
2	Comedy (Red)	Horror (Yellow)	Action (Blue)	Romance (Green)
Frequency	7	9	8	1
Relative Freq.				

	A	B	C	D
3	Math (Red)	English (Yellow)	Science (Blue)	Health (Green)
Frequency	4	6	11	4
Relative Freq.				

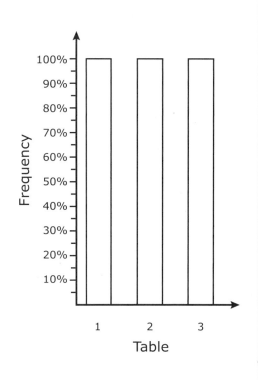

Unit 37 Homework

Standard 6.12(D) – Readiness

Use the information to answer questions 1–3.

A group of students is surveyed to determine their music genre preference. The results are shown below.

Music Genre	Students
Alternative	||||
Country	||||	
Hip-Hop/Rap	|||| |||| ||
Pop	|||| |||| ||||	
Rock	|||| ||||

1 Complete the table below.

Music Genre	Frequency	Relative Frequency
Alternative		
Country		
Hip-Hop/Rap		
Pop		
Rock		

2 Which music genre represents the mode of the data?

_____

3 Complete the following statements.

The music genre _____ was selected twice as much as the genre _____.

The music genre _____ was selected about one-third as much as the genre _____.

Use the information to answer questions 4–6.

The percent bar graph shows the results of a survey conducted at a local ice cream parlor about their customers' favorite flavors of ice cream.

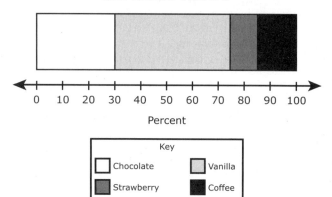

4 If a total of 60 customers were surveyed, complete the table below.

Ice Cream Flavor	Frequency	Relative Frequency
Chocolate		
Vanilla		
Strawberry		
Coffee		

5 Which ice cream flavor represents the mode of the data?

_____

6 _____-flavored ice cream was selected three times as often as _____-flavored ice cream.

Connections

1. Search the Internet for different representations of percent bar graphs. Identify at least 3 different ways a percent graph is displayed. Print out the examples and summarize the similarities and the differences between the examples.

2. Write a survey question to ask fellow classmates. Use the data you collect to create a relative frequency table and a percent bar graph. Display the information on a poster and share with the class.

Name _____

Standard 6.13(A) – Readiness

Unit 38 Introduction

1 Adrian receives weekly grades for his math homework assignments. Each week is based on a 100-point scale. The stem-and-leaf plot shows Adrian's weekly homework grades for the first two six weeks.

Adrian's Homework Grades

Stem	Leaves
7	3 8 8
8	0 4 6 9
9	0 2 5 5
10	0

7 | 5   represents 75

What fraction of Adrian's homework grades were 80 or above?

_____

Fill in the blank to make the sentence true.

Of Adrian's math homework grades, $33\frac{1}{3}\%$ were _____.

2 The Stingrays soccer team records the number of goals scored by each teammate during the season, as shown in the dot plot below.

What percent of the teammates scored more than 4 goals for the season?

_____

How many points did the Stingrays score during the season?

_____

3 Tran researches the average life span of different types of birds and records the data in the histogram shown below.

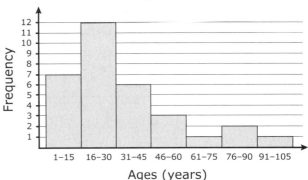

How many types of birds did Tran include in his research?

_____

What percent of the birds in Tran's research have an average life span of more than 60 years?

_____

4 Dakota gathers data on the number of students who attended art club meetings in the first semester. She created a box plot of the data, as shown below.

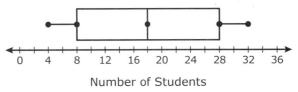

What percent of the meetings show attendance between 8 and 28 students?

_____

One-fourth of the art club meetings have an attendance between _____.

Unit 38 Guided Practice

Standard 6.13(A) – Readiness

Use the graphs to answer questions 1 and 2.

Mr. Dean and Mr. Mack each make a box plot to represent the grades from the same unit test for their first-period classes.

Mr. Dean's Unit Test Grades

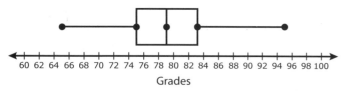

Mr. Mack's Unit Test Grades

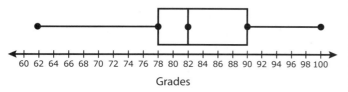

Use the graph to answer questions 3 and 4.

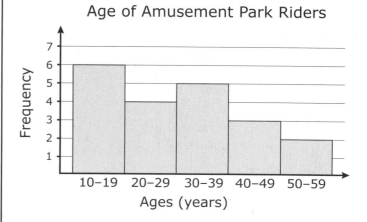

1. The number of students in Mr. Mack's first period that scored between 78 and 90 on the unit test represent what percent of his students?

   Ⓐ 25%

   Ⓑ 50%

   Ⓒ 75%

   Ⓓ 100%

2. Which of the following does NOT support the data as shown in the box plots?

   Ⓕ One-half of the test grades for Mr. Dean's class range from 75 to 83.

   Ⓖ One-fourth of the test scores for Mr. Dean's class range from 65 to 75.

   Ⓗ The median of Mr. Mack's class grades is 82.

   Ⓙ One-fourth of the test grades for Mr. Mack's class range from 89 to 100.

3. Which of the following statements is supported by the histogram?

   Ⓐ Of the 20 riders represented in the graph, over 50% are 30 years of age or older.

   Ⓑ One-fourth of the riders represented in the graph are younger than 40 years old.

   Ⓒ Half the riders represented in the graph are between the ages of 10 and 29.

   Ⓓ Six percent of the riders represented in the graph are between the ages of 10 and 19.

4. What value, in years, represents the spread of the data?

   Record your answer and fill in the bubbles on the grid below. Be sure to use the correct place value.

Standard 6.13(A) – Readiness

Unit 38 Independent Practice

Use the graph to answer questions 1 and 2.

Salvador represents the heights, in inches, of the male students in his class using the stem-and-leaf plot below.

Heights of Male Students (in.)

Stem	Leaves
5	8
6	0 2 4 5 5 7 9 9
7	3

5 | 8   represents 58

1. What percent of the male students in Salvador's class are taller than 5 feet 5 inches?

   Ⓐ 100%

   Ⓑ 40%

   Ⓒ 60%

   Ⓓ 50%

2. Which of the following statements is supported by the information shown in the graph?

   Ⓕ There are no male students in Salvador's class taller than 6 feet.

   Ⓖ There is only one male student in Salvador's class that is 5 feet 8 inches tall.

   Ⓗ There are eight male students in Salvador's class whose height ranges from 5 feet to 6 feet tall.

   Ⓙ Twenty percent of the male students in Salvador's class are shorter than 5 feet.

Use the graph to answer questions 3 and 4.

Mrs. Raabe's homeroom class conducted a survey to determine the number of pets in each student's household. They organized the information in a dot plot.

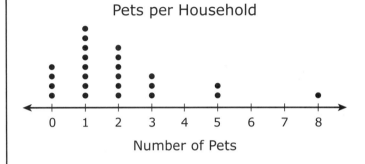

Pets per Household

3. Which of the following statements is NOT supported by the data in the dot plot?

   Ⓐ One-fourth of the students in Mrs. Raabe's class have exactly two pets.

   Ⓑ Fifty percent of the students in Mrs. Raabe's class have more than one pet.

   Ⓒ One-third of the students in Mrs. Raabe's class have just one pet.

   Ⓓ Twenty percent of the students in Mrs. Raabe's class have more than two pets.

4. The total number of pets owned by students in Mrs. Raabe's class is —

   Ⓕ greater than 45, but fewer than 50

   Ⓖ greater than 5, but fewer than 10

   Ⓗ greater than 30, but fewer than 35

   Ⓙ greater than 50, but fewer than 55

Unit 38 Assessment

Name _____

Standard 6.13(A) – Readiness

1 Ginger practices daily for the upcoming cheerleader tryouts. She displays her practice times in the dot plot shown below.

Ginger's Practice Record

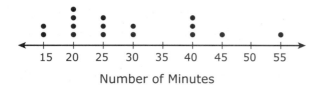

Number of Minutes

How many days did Ginger practice for longer than a half hour?

Ⓐ 2 days

Ⓑ 7 days

Ⓒ 9 days

Ⓓ 5 days

2 Mrs. Beck shops from the clearance rack of dresses. The stem-and-leaf plot represents the sale prices of dresses in Mrs. Beck's size.

Clearance Dress Prices

Stem	Leaves
1	3 5 6 8
2	0 1 4 4 5 5 9
3	3 3 4 5 7
4	0 2 2 6
5	4 9
6	5 5
7	0

1 | 3   represents 13

Which statement is true?

Ⓕ Twenty percent of the dresses cost more than $50.

Ⓖ One-fifth of the dresses range in price from $30 to $40.

Ⓗ Fifteen percent of the dresses cost less than $20.

Ⓙ Four percent of the dresses cost $65.

3 Lila records the weights, in pounds, of 20 packages delivered on Monday. The box plot below represents the data she recorded.

Package Weight

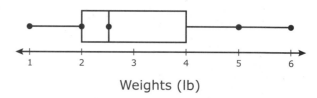

Weights (lb)

What percent of the packages had a weight in the interval from 1 to 2 pounds?

Ⓐ 20%

Ⓑ 50%

Ⓒ 25%

Ⓓ 15%

4 The city library offers events for children during the summer. The histogram below represents the ages of children who attend the first event of the summer.

Summer Kick-Off Event

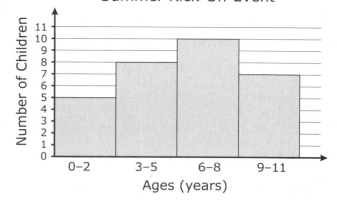

Ages (years)

What fraction of the children in attendance at the library event were between the ages of 6 and 8?

Ⓕ $\frac{1}{4}$

Ⓖ $\frac{1}{3}$

Ⓗ $\frac{1}{2}$

Ⓙ $\frac{1}{5}$

Name _____

Standard 6.13(A) – Readiness

Unit 38 Critical Thinking

1 Two histograms containing the same number of data points are shown.

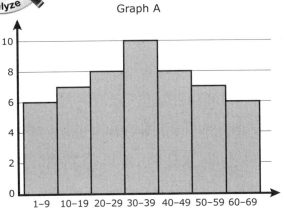

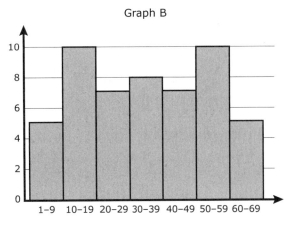

In which interval would the mean and median be located for each histogram? Explain how you know.

_____
_____

How many modes does each graph show? Explain.

_____

2 The two box plots show the number of questions answered correctly by different students in two different classes.

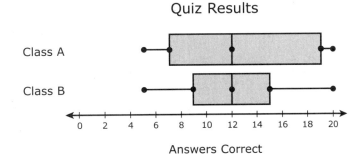

Describe the distribution around the center for the number of correct answers in the two classes.

_____
_____

Which class do you think performed better on the quiz? Justify your answer with data from the box plots.

_____
_____

Unit 38 Journal/Vocabulary Activity

Name _____

Standard 6.13(A) – Readiness

Journal

Explain why the median of a data set is not always in the center of the box in a box plot.

_____

_____

Vocabulary Activity

Use the clues below to locate 15 vocabulary terms studied in this unit. Challenge: Find a 16th term.

```
e u y b t k c m t s r d x o z m j r
z g p s c o r a y w a a m f e s e u
s h n w t o l m o k n e j a r t f f
m k s a f e m p o e g r n w n o l b
s b e i r e m s x x e p z e b l l k
c i n w t e i a x o c s c o d p a c
r u y r e x l b n y b f e l i t r a
h u i o p d a i s d o o s t x o f b
i c h v e p r h t s l s d z f d g a
s z y n d l f c e r k e o a m i z n
t w j f h b y r c y a r a n o o b x
o j n e f x u k j n k u r f i l r j
g f c d j s i i u q h s q t p f q p
r s a x a e v f b r u a e r s l f y
a y t e l i t r a u q e t c e d o m
m h m s i f p w x q j m y s h t k t
d a t a d i s t r i b u t i o n n j
q x w h a e m e d i a n i w o g p i
```

1. Uses a rectangle to represent the middle 50% of a data set
2. Uses dots to show frequency of data on a number line
3. Organizes data from least to greatest using place value to group data
4. A graph in which the labels for each bar are numerical intervals
5. The calculated center of a data set when the data is uniform or symmetric
6. The middle number of a set of numbers when the numbers are ordered
7. The most frequent value in a set of numbers
8. The difference between the largest and smallest values in a data set
9. The difference between the values of the upper and lower quartiles
10. Measures used to describe data, including range and interquartile range
11. Measures used to describe data, including mean and median
12. A shape of data in which the data is equally distributed
13. A shape of data in which the majority of data is contained to one side or the other
14. A shape of data in which the data is equally divided around the center of the data
15. Divides an ordered set of data into four equivalent, or close to equivalent, groups

Name _____

Standard 6.13(A) – Readiness

Unit 38 Motivation Station

Statistical Statements

Play *Statistical Statements* with a partner. Each pair of players needs a game board and a number cube. Each player needs a pencil. Player 1 begins by rolling the number cube to determine a statement. The player then analyzes the graphs to determine which best matches the statement rolled, and gives a justification. If both players agree, player 1 initials both the statement and the graph and play passes to player 2. If the players disagree, player 1 loses a turn. The game continues until each graph has been matched with a correct statement. The winner is the player with more statements and graphs initialed.

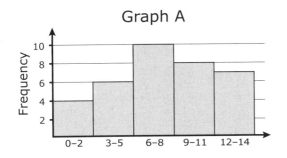

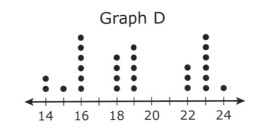

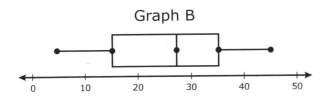

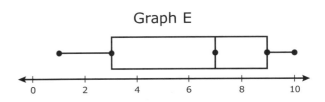

Graph C

Stem	Leaves
0	2 3 3 5 6
1	1 1 5
2	1 1 2

0 | 2   represents 2

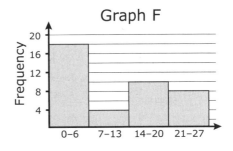

 There are 2 modes shown in this graph.

 Less than 25% of the data in this graph is greater than 20.

More than 50% of the data in this graph is greater than 25.

The mean is less than the median in this graph.

The spread of the data in this graph is 1.5 times the interquartile range.

The frequency in this graph is equal to 35.

Unit 38 Homework

Name _____

Standard 6.13(A) – Readiness

1 The box plot shows the amounts spent each month by 120 students on school lunches.

School Lunch Spending

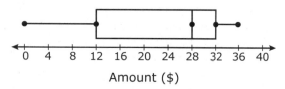

Amount ($)

How many students spent $28 or less each month on school lunches?

Is the spread for the 60 students who spent the most money on lunches greater than or less than the spread for the 60 students who spent the least? Explain how you know.

_____
_____
_____

2 The nurse at Cumberland School recorded the weights of students who visited the clinic on Friday.

Weights of Students (lb)

Stem	Leaves
5	7 9
6	3 5 7 8 8
7	0 2 2 2 4 6 9
8	0

5 | 7   represents 57

Write *true* or *false* beside each statement.

Fewer than 50% of the students weigh less than 70 pounds. _____

The mode of the students' weights is 72 pounds. _____

The median of the students' weights is 68 pounds. _____

3 The histogram shows the results of a survey about the number of hours students watch television on the weekend.

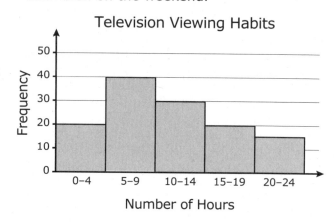

How many students participated in the survey?

_____

True or false? Over 50% of students watch between 5 and 14 hours of television on the weekend. Explain.

_____
_____

4 The dot plot shows the number of books read by students in Mrs. Young's class over summer break.

Summer Break Reading

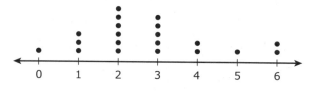

Number of Books Read

How many books did the students read over summer break? Explain how you know.

_____
_____
_____

Connections

Locate different graph types in a science and/or social studies textbook. Sketch the graphs on a sheet of paper and describe the information in the graphs.

Name _____

Standard 6.13(B) – Supporting

Unit 39 Introduction

1 What is statistical data?

_____
_____
_____

2 Below are two examples of situations that yield variability (statistical questions).

- What is the price of a new car?
- How many potatoes should I buy to make French fries for the class?

Create two questions of your own in which the data yields variability.

_____
_____
_____
_____

3 Below are two examples of situations that do not yield variability (non-statistical questions).

- What is the date today?
- How many wheels are on a tricycle?

Create two questions of your own in which the data does NOT yield variability.

_____
_____
_____
_____

4 Identify each of the following as *statistical* or *non-statistical*. If the question is statistical, explain how you know. If the question is non-statistical, rewrite the question so that it yields variability.

a. How many library books are in each student's backpack?

_____
_____
_____

b. How many library books were checked out last week at your school?

_____
_____
_____

c. How much would you owe your school library if you returned your book 5 days late?

_____
_____
_____

d. How many students pay fines to the library each day at your school?

_____
_____
_____

e. How old are the students who visit the library at your school?

_____
_____
_____

Unit 39 Guided Practice

Standard 6.13(B) – Supporting

1 Which situation does NOT yield variability?

Ⓐ How many days are in June?

Ⓑ How many students are in a sixth-grade classroom?

Ⓒ How many texts do you receive each month?

Ⓓ How many babies are born each day at Mother Frances Hospital?

2 For extra practice, Yazmine needs to identify an example and non-example of a statistical question. Which of the following should Yazmine choose?

Ⓕ With variability: How many quarts are in two gallons?

   Without variability: How many different flavors of lollipops are in the bag?

Ⓖ With variability: What are the heights of the students in your class?

   Without variability: How many hours per week do you spend working on homework?

Ⓗ With variability: What grade did you make on your last math test?

   Without variability: What is the cost of a bag of Doritos® from the vending machine at your school?

Ⓙ With variability: How many hours do you sleep each night?

   Without variability: How many consonants are in the word "forsythia"?

3 Gary's Pediatric Clinic is promoting health and fitness. They have asked an advertising firm to create questions that would yield variability within the results. Which question would NOT allow for any variability?

Ⓐ How many weeks per year do students exercise at school?

Ⓑ On average, how many days per week do students exercise at school?

Ⓒ How many minutes does your school schedule for P.E. each week?

Ⓓ On average, how many hours per week do students exercise at school?

4 Which of the following questions allow for variability?

   I. How many students visited the nurse on Monday, September 15, 2014, at your school?

   II. How many free throws can you make in one minute?

   III. How much money do doctors earn per year?

   IV. How tall are the players on the Harlem Globetrotters?

Ⓕ II and III only

Ⓖ II, III, and IV only

Ⓗ I, II, and IV only

Ⓙ I, II, III, and IV

Name _____

Standard 6.13(B) – Supporting

Unit 39 Independent Practice

1. Mrs. Lebowitz uses statistical questions to generate data about the students in her class. For which of the following survey questions would Mrs. Lebowitz NOT receive variable results with her class of sixth-grade students?

    Ⓐ How many pets does each student have?

    Ⓑ How old, in years, are the students in the class?

    Ⓒ In what grade are the students?

    Ⓓ What is the height, in inches, of each student in the class?

2. Tyrese attends dental school and is studying the effects of good dental hygiene. He must generate a heading for his research that represents statistical data. Which of the following would allow Tyrese to research data with variability?

    Ⓕ How many preventive care visits have you had in the last 12 months?

    Ⓖ How much money does your family spend per year at the dentist?

    Ⓗ How many times did you floss your teeth last week?

    Ⓙ How many cavities have you had filled in the last 5 years?

3. Which of the following situations does NOT yield variability?

    I. How many grams of sugar are in a 12-ounce can of Dr. Pepper®?

    II. How much sugar is used to make a gallon of punch?

    III. How many grams of sugar does a teenager consume each day?

    IV. How much sugar is in an 8-ounce glass of chocolate milk?

    Ⓐ I only

    Ⓑ I and II only

    Ⓒ II and III only

    Ⓓ I and IV only

4. The local Boys and Girls Club gathers data about youth programs in the community. Which of the following would allow for variability in the results?

    Ⓕ What year was the Boys and Girls Club program first brought to your town?

    Ⓖ How many schools offer Boys and Girls Club programs in your town?

    Ⓗ How many schools do not have an after-school program in your town?

    Ⓙ What age child attends Boys and Girls Club programs in your town?

Unit 39 Assessment

Name _____

Standard 6.13(B) – Supporting

1 Mr. Williams surveys the new student in his class using the questions below.

Question 1: How many pets do you own?

Question 2: How many times each month do you purchase a lunch in the cafeteria?

Question 3: How many hours did you spend reading yesterday after school?

Question 4: How many absences from school did you have last week?

Which statement is NOT true about the questions asked in Mr. Williams' class?

Ⓐ Question 1 is non-statistical because it does not allow variability in responses.

Ⓑ Question 2 is statistical because the responses will vary each month.

Ⓒ Question 3 is non-statistical because there is only one response.

Ⓓ Question 4 is statistical because the responses will vary each week.

2 The local humane society gathers information regarding the number of pet adoptions. Which of the following questions could show variability within the results?

Ⓕ How many pets were adopted on Friday, October 3, 2014?

Ⓖ What is the difference between the number of dog adoptions last year and the number of cat adoptions last year?

Ⓗ What are the ages of the pets adopted yesterday?

Ⓙ How many volunteers helped with the adoption clinic at the mall last weekend?

3 Which situation does NOT yield variability?

Ⓐ How many miles do you travel in your car each day?

Ⓑ How many days are in a leap year?

Ⓒ How many students attend the Six Flags® band trip each year?

Ⓓ What are the ages of the students taking Spanish at your school?

4 Which of the following are statistical questions?

I. How many of Mrs. Malone's students earned an A the first nine weeks grading period of math?

II. How many students missed question #5 on Mr. Blevins' science homework last night?

III. How many miles does Kendrick run each week in athletics?

IV. What was Ethan's highest average on his report card for the entire school year?

Ⓕ III only

Ⓖ III and IV only

Ⓗ I, III, and IV only

Ⓙ I, II, and III only

Name _____

Standard 6.13(B) – Supporting

Unit 39 Critical Thinking

1. The following claim appears in a newspaper advertisement for a furniture store.

> **Our prices are the lowest in town...Guaranteed!**

Is it possible to use statistics to prove this claim true or false? If you answer *yes*, write a statistical question that could be used to gather data to prove or disprove the claim. If you answer *no*, explain why you do not believe it is possible.

_____
_____
_____
_____

2. Design a mini-statistical project. Begin by identifying a topic you or people your age might want to know more about.

_____

Next, write a statistical question involving the topic selected. The question will need to yield variability in the data collected.

_____

Describe the variability that is expected in the data, based on the question above.

_____

Describe the people who will be asked to answer the question (population), and determine how many people will be surveyed to ensure good data (sample size).

_____

Make a prediction about the results of the data that will be gathered. What type of generalizations will be revealed through the survey and why?

_____
_____
_____

Unit 39 Journal/Vocabulary Activity

Name _____

Standard 6.13(B) – Supporting

Journal

Explain, in your own words, why a single number cannot yield statistical data.

_____
_____
_____

Vocabulary Activity

Complete each graphic organizer below.

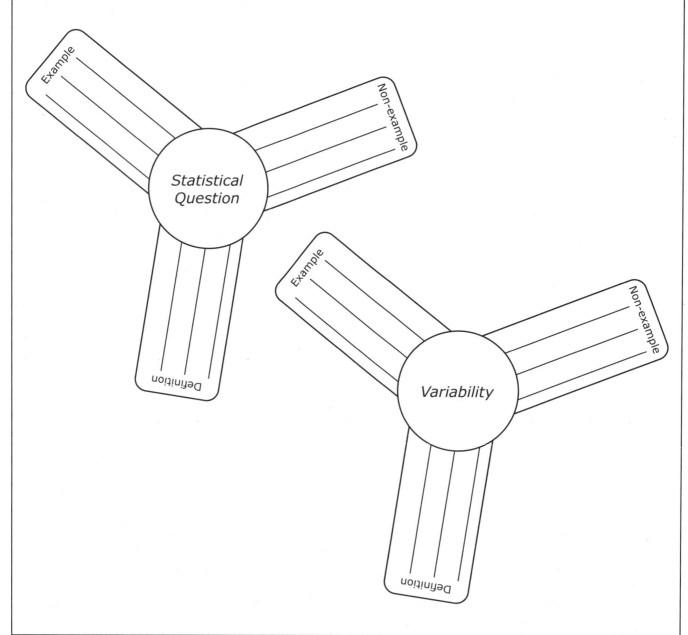

Name _____

Standard 6.13(B) – Supporting

Unit 39 Motivation Station

Steve's Statistical Dilemma

Complete *Steve's Statistical Dilemma* individually. Help Steve find his way across town to the school, so he does not miss his math lesson on statistics! The path he uses may only contain questions that yield variability (statistical questions). Show the path Steve will use by shading the boxes that contain statistical questions.

How tall are professional basketball players?	How many cookies does a bakery produce in a week?	How old was John Adams when he became president?	How many games did the football team win last season?
How many state names begin with the letter *A*?	How many hours a night do sixth-grade students study?	How many movies did Sally see at the theater last month?	How many days was the high temperature over 100°F in 2012?
How tall was Shaquille O'Neal when he was 18 years old?	How many Americans work more than 40 hours per week?	What is the daily high temperature in Dallas in June?	How many desks are in Mrs. Ellis' classroom?
How many feet are in one mile?	What was Hank Aaron's batting average during his final season?	What size shoe does an 11-year-old wear?	How much water does a family use in a month?

©2014 mentoringminds.com motivationmath™ LEVEL 6 ILLEGAL TO COPY 317

Unit 39 Homework

Standard 6.13(B) – Supporting

1. For each of the following questions, write *yields variability* or *does not yield variability*, as applicable.

   a. What did Grace score on her last math quiz?

   b. What did the sixth-grade students score on the last math quiz?

   c. In a jar filled with different buttons, what is the size of a button?

   d. How many buttons are in the jar?

2. For each of the following, rewrite the question to be a statistical question.

   a. How old are you?

   b. What was the difference in rainfall amounts in Austin in 2012 and 2013?

   c. What was the highest score on the last math benchmark assessment?

   d. How many text messages did you send yesterday?

3. Jana and Raeane work together on a statistics project for their math class. The first step in the project is to write a statistical question that will yield data with variability. The question Jana and Raeane write is:

   How often do students from this school go to the movies?

   Will the question Jana and Raeane wrote yield variability? Explain.

4. Classify each question as *statistical* or *non-statistical*.

   a. How many sodas do students in this school drink in one month?

   b. How many M&M'S® are in a bag?

   c. What was your grade in math last semester?

   d. How much rain does Brazil receive in a year?

   e. How much does a new television cost?

   f. How much money did Jill receive on her birthday?

Connections

1. Create a poster with 4 examples of statistical data from newspapers, magazines, or printed from the Internet. In the center of the poster, write *Statistical Data*, and paste the examples in the corners.

2. Write your own statistical question and conduct a survey of students at your school. Use the data collected to create a display with a written explanation on the back about why your question is statistical, what the shape and spread of the data distribution tell about the data collected, etc.

Unit 40: Standards 6.14(A) – Supporting, 6.14(C) – Supporting

Unit 41: Standards 6.14(B) – Supporting, 6.14(D), 6.14(E) – Supporting, 6.14(F) – Supporting

Unit 42: Standards 6.14(G) – Supporting, 6.14(H) – Supporting

Name _____

Standards 6.14(A) – Supporting, 6.14(C) – Supporting

Unit 40 Vocabulary Activity

The following terms and phrases will be used throughout this unit.

ATM	credit union	financial institution	service charge (fee)
balance	debit card	minimum balance	transaction
check register	deposit	online banking	transfer
checking account	direct deposit	overdraft	withdrawal

Four-in-a-Row

Play *Four-in-a-Row* with a partner. Player 1 begins by choosing a term and giving the definition. Player 2 verifies the definition, using the glossary. If correct, player 1 initials the space containing the term. The players then switch roles and continue with the next term chosen by player 2. The first player to initial four boxes in a row horizontally, vertically, or diagonally is the winner.

direct deposit	online banking	check register	withdrawal
ATM	transaction	minimum balance	transfer
service charge (fee)	checking account	financial institution	overdraft
balance	deposit	debit card	credit union

©2014 mentoringminds.com   motivation**math**™ LEVEL 6   ILLEGAL TO COPY   321

Unit 40 Practice

Standards 6.14(A) – Supporting, 6.14(C) – Supporting

Raquel just started her first job and needs to open a checking account. She and her mother begin researching banks online to determine which bank Raquel should use. They narrow the list to two financial institutions in their town: CAN Credit Union and SBA Bank. The charts below compare the different checking account options at the two banks.

CAN Credit Union

	Advantage	Platinum	Platinum Plus
Minimum Opening Deposit	$50	$50	$50
Monthly Service Charge	None	$5 unless minimum balance of $1,000 maintained	None
Debit Card	Yes	Yes	Yes, with $0.10 fee per transaction
Monthly Interest Earned	Paid monthly on balances of $1,000 or more	None	None
Online Banking	Yes	Yes	Yes

SBA Bank

	NOW Checking	Classic Checking	Spirit Checking
Minimum Opening Deposit	$50	$50	$50
Monthly Service Charge	$7.50 unless minimum balance of $1,000 maintained	$2, none with e-statement	None
Debit Card	Yes	Yes	Yes
Monthly Interest Earned	Paid monthly depending on daily balance	None	None
Online Banking	Yes	Yes	Yes

Use the information in the charts to answer the following questions.

1. Explain the differences between the Platinum and the Platinum Plus accounts at CAN Credit Union.

2. Explain what might be appealing to Raquel about the NOW Checking account offered at SBA Bank. Do you think this is a good choice for Raquel? Explain.

3. Which bank and account would you suggest for Raquel? Explain.

Name _____

Standards 6.14(A) – Supporting, 6.14(C) – Supporting

Unit 40 Practice

Raquel chooses SBA Bank's Spirit Checking account. She opens the account with an initial deposit of $50 she received for her birthday. Raquel chooses to have overdraft protection on her account. This means if she does not have the funds to cover a transaction, the bank will pay the debit and charge her a $30 overdraft item fee. The check register below shows the transactions Raquel made during the first month her account was open. Some of the amounts are missing.

Check Number	Date	Transaction Description	(−) Payment/ Debit		(−) Fees		(+) Deposit/ Credit		Balance	
	1/5	Initial deposit					50	00	50	00
	1/8	Paycheck					121	80	171	80
101	1/9	Mom – cell phone payment	50	00						
	1/15	Paycheck					152	25	274	05
102	1/16	Car insurance							229	05
	1/18	Non-SBA ATM - Cash	30	00					199	05
	1/18	ATM fee							197	05
	1/22	Paycheck					121	80	318	85
	1/22	Clothes Barn	53	46						
	1/28	Burger-U	9	52					255	87
	1/29	Paycheck							377	67
103	1/29	SBA – Car payment	110	63						
	1/30	Jumpin' Tacos	14	32					252	72
	1/31	Transfer to savings	25	00					227	72

4 What was the balance in Raquel's account on January 9 after she paid her cell phone payment to her mom?

_____

5 For what amount did Raquel write a check to pay for her car insurance?

_____

6 How much was the ATM fee charged to Raquel's account on January 18?

_____

Explain what this fee is for.

_____

7 What amount did Raquel deposit into her account on January 29?

_____

8 Could Raquel have chosen to make her car payment on January 14 instead of January 29?

_____

Explain your answer.

_____

Unit 40 Assessment

Standards 6.14(A) – Supporting, 6.14(C) – Supporting

Use the information below to answer questions 1 and 2.

Trinity Credit Union	Fairfield Bank & Trust
*$9.95 per month, save $5 with direct deposit *no non-network ATM fees *$25 overdraft item fee per item *debit card offered *free online banking	*$6.50, free with a minimum balance of $200 *$2.50 ATM fee per non-network transaction *debit card offered *$30 overdraft item fee per item

1. Silas has a checking account at Fairfield Bank & Trust. On the morning of May 17 he had a balance of $168.42. On the morning of May 19, he had a balance of $59.79. Which of the following transactions could NOT be the reason for Silas' new balance?

   Ⓐ Silas transferred $108.63 from his checking account to his savings account.

   Ⓑ Silas withdrew $106.13 from a non-network ATM.

   Ⓒ Silas wrote a check for $58.63 and then deposited $50 into his checking account.

   Ⓓ On May 17, Silas paid a $175 bill, and on May 18, he deposited $96.37.

2. Mr. Milstein banks at Trinity Credit Union. His account has a balance of $331.24 at the beginning of the month. He visits a non-network ATM to withdraw $275. That night, a check Mr. Milstein wrote for $78.68 clears his bank. The next day, Mr. Milstein's paycheck for $512.45 is deposited to his account via direct deposit. Which amount best describes the fees Mr. Milstein has incurred so far this month?

   Ⓕ $25

   Ⓖ $29.95

   Ⓗ $34.95

   Ⓙ $30

3. Look at the incomplete check register shown below.

Check Number	Date	Transaction Description	(−) Payment/Debit		(−) Fees		(+) Deposit/Credit		Balance	
	10/2	Balance							164	35
	10/4	Transfer to savings	55	00						
108	10/5	Sally's Spa	71	45						
	10/6	Paycheck					201	98		
	10/8	ATM (non-network)	45	00	2	95				

What is the balance in the account after these transactions?

Record your answer and fill in the bubbles on the grid below. Be sure to use correct place value.

Name _____

Standards 6.14(A) – Supporting, 6.14(C) – Supporting

Unit 40 Motivation Station

Balancing Act

Complete *Balancing Act* individually. Read each transaction description below. Use the information to complete Raquel's check register for the month of February. Raquel's bank assesses an overdraft fee of $30 for each transaction that results in a negative balance. Raquel is charged an ATM fee of $2 if she uses a non-SBA ATM for cash withdrawals.

Transaction descriptions:

1. Uses debit card to purchase a formal gown at Francesca's Formal Fashions for $162.38 on February 2
2. Paycheck in the amount of $152.25 deposited on February 5
3. Writes check #104 for $50 to Mom for cell phone payment on February 9
4. Uses debit card to pay for electronic music download on iTunes® for $10.81 on February 10
5. Paycheck in the amount of $60.90 deposited on February 12
6. Writes check #105 for $45 for car insurance on February 15
7. Uses SBA ATM to withdraw $100 cash for weekend trip on February 17
8. Writes check #106 for $75.76 to purchase accessories for formal at Accessory Hut on February 17
9. Paycheck in the amount of $121.80 deposited on February 19
10. Transfers $32.46 to Dad's account for oil change on February 22
11. Uses debit card to purchase lunch at Chicken Shack for $7.87 on February 23

Check Number	Date	Transaction Description	(−) Payment/ Debit		(−) Fees		(+) Deposit/ Credit		Balance	
	2/1	Balance							227	72

Unit 40 Homework/Connections

Standards 6.14(A) – Supporting, 6.14(C) – Supporting

Use the information below to answer questions 1–5.

	Better Bank for All	Consumer Friends Bank	WOW Bank
Monthly Fee	$4, unless minimum balance of $500 maintained	$5, free with direct deposit	$2, free with e-statement
Non-network ATM Fees	$3.00 per transaction	$1.50 per transaction	$2.00 per transaction
Overdraft Item Fee	$35 per item	$25 per item	$25 per item

1. If you maintain a balance of $500, choose to have your paycheck directly deposited to your account, and opt for electronic bank statements, which financial institution is the most economical? Explain your reasoning.

2. Bridgett has a balance of $128.59 in her checking account at Better Bank for All. She transfers $75 to her savings account. She writes a check for $54.74 to pay her credit card bill. If there are no other transactions, what will be the balance in Bridgett's account once her check clears?

3. Mrs. Morris has an account at WOW Bank with a current balance of $570.43. She uses a non-network ATM to withdraw $450. What is Mrs. Morris' new balance?

4. Mr. Bryant's checking account is with Consumer Friends Bank. He does not have direct deposit for his weekly paychecks, but he does receive his bank statements electronically. This month he has withdrawn money using a non-network ATM three times and a network ATM five times. Also this month, he had one item paid using overdraft protection. How much did Mr. Bryant pay in fees for the month?

5. Quinn had a balance of $326.12 in his checking account at WOW Bank on September 1. On September 2, he visited a non-network ATM to make a withdrawal. On September 3, he wrote a check to West Water Supply, and on September 4, he made a deposit. After those 3 transactions, Quinn had a new balance of $214.20. Complete the check register below to show the transaction amounts Quinn might have made.

Check Number	Date	Transaction Description	(−) Payment/Debit	(−) Fees	(+) Deposit/Credit	Balance	
	9/1	Balance				326	12

Connections

Research opening a checking account at two local financial institutions. Create a table comparing each financial institution's costs for opening and maintaining the account, debit card usage, and any other associated fees or benefits. Then explain which option you feel is best. Share your findings with the class.

Name _____

Standards 6.14(B) – Supporting, 6.14(D), 6.14(E) – Supporting, 6.14(F) – Supporting

Unit 41 Vocabulary Activity

The following terms and phrases will be used throughout this unit.

annual fee	credit limit	grace period
annual percentage rate (APR)	credit report	interest
borrower	credit score	late fee
collection account	debit card	lender
credit card	delinquency	minimum payment

As you read the paragraphs below, fill in the blanks with terms from the vocabulary list. A term may be used more than once.

When making purchases, there are typically four methods of payment: cash, check, (1)_____, or (2)_____. When using a (3)_____, the money is automatically withdrawn from a bank account. When using a (4)_____, the money is "borrowed" from the company giving the credit, the (5)_____, and repaid with interest. The (6)_____ is the amount charged for "using" the (7)_____'s money and is calculated as a percentage of the amount charged. This percentage is called the APR or (8)_____. There is typically a (9)_____ before the interest is added to the account, depending on the terms of using the card. Some card companies also charge an (10)_____ to have their card, adding to the cost of using credit.

When credit is used, the information about the borrower's use of credit is recorded in a (11)_____. Making at least a (12)_____ by the given due date every month, and not charging so much on the credit card that the balance stays close to the (13)_____ will give the borrower a positive credit history. Failing to make payments by the given due date not only results in a negative credit history by adding a delinquency to the report, but may also cause a (14)_____ to be added to the account. Every (15)_____ reported on an account lowers a borrower's credit score. A (16)_____ is a number that a future (17)_____ uses to determine a (18)_____'s likelihood to repay a loan or credit card. In addition to a delinquency, having an account's status changed to collection account will also lower the (19)_____. A (20)_____ is an account that a company has turned over to another company to attempt to collect the debt because the (21)_____ has refused to pay what is owed.

Unit 41 Practice

Standards 6.14(B) – Supporting, 6.14(D), 6.14(E) – Supporting, 6.14(F) – Supporting

Tomas turned 18 over a year ago and was approved for his first credit card. Until then, he used his debit card to make purchases, and the money came directly from his checking account. His credit card has a $300 credit limit, and Tomas uses the credit card frequently. Tomas is trying to purchase his first car, but the bank has denied his loan application. His father helps him pull his credit report and credit score from one of the credit bureaus to determine why his application was denied. The graphic below shows part of Tomas' credit report.

Personal Information

Name: Tomas Rojas	SSN: XXX-XX-3289
Other Names: None	Date of Birth: April 2, 1994
Report Number: 35884219	Telephone Number: XXX-XXX-0313
Report Date: August 20, 2013	
Current Address: 1248 Wild Oak Drive, New Town, Texas 75788	Previous Address: None
Employment Data Reported Employer Name: Burger Shack Date Reported: October 2010	Location: New Town, Texas Date Hired: September 2010

Potentially Negative Items

Credit Items		
Alpha Credit Company		Status: Open/Late
Date Opened: 06/2012	Type: Charge Card	Credit Limit/Original Amount: $300
Reported Since: 07/2012	Terms: NA	High Balance: $320
Date of Status: 08/2013	Monthly Payment: $25	Recent Balance: $320 as of 07/2013
Last Reported: 07/2013	Responsibility: Individual	Recent Payment: $0
Account History: 30 days past due as of 06/2013; 60 days past due as of 07/2013		

Credit Score Information

Your current score:

609

Credit Rating: Poor

---

1 After reviewing Tomas' credit report, what reason do you think the bank could give for denying his loan application? Explain.

_____

_____

2 How long will the negative information remain on Tomas' credit report? _____

3 Tomas is confused as to why the bank pulled his credit report. He is young and he only has one credit card. Why do banks and other lenders look at a person's credit history before allowing them to borrow money or open an account? How does looking at a credit report help a borrower?

_____

_____

Name _____

Standards 6.14(B) – Supporting, 6.14(D), 6.14(E) – Supporting, 6.14(F) – Supporting

Unit 41 Practice

Tomas works for 6 months to pay off his credit card account and does not use the card at all during that time. He also saves $500 to pay toward his car. When he goes to the bank to apply for a car loan again, his credit report shows the following.

Personal Information	
Name: Tomas Rojas Other Names: None Report Number: 36771245 Report Date: February 28, 2014	SSN: XXX-XX-3289 Date of Birth: April 2, 1994 Telephone Number: XXX-XXX-0313
Current Address: 1248 Wild Oak Drive, New Town, Texas 75788	Previous Address: None
Employment Data Reported Employer Name: Burger Shack Date Reported: October 2010	Location: New Town, Texas Date Hired: September 2010

Potentially Negative Items		
Credit Items			
Alpha Credit Company		Status: Open/Late
Date Opened: 06/2012 Reported Since: 07/2012 Date of Status: 02/2014 Last Reported: 02/2014	Type: Charge Card Terms: NA Monthly Payment: $25 Responsibility: Individual	Credit Limit/Original Amount: $300 High Balance: $320 Recent Balance: $0 as of 02/2014 Recent Payment: $35
Account History: 30 days past due as of 06/2013; 60 days past due as of 07/2013		

Credit Score Information	
Your current score: **640**	Credit Rating: Fair

4 Why is Tomas' account still shown under *Potentially Negative Items*?

_____

_____

5 How has Tomas' credit rating changed from 6 months ago? What could have caused the change?

_____

_____

6 Tomas has been paying for his purchases using his debit card. Why is this fact not reported as positive on his credit report?

_____

_____

7 Tomas is approved by the bank for his car loan. What actions should Tomas take to ensure his credit rating continues to improve?

_____

Unit 41 Assessment

Standards 6.14(B) – Supporting, 6.14(D), 6.14(E) – Supporting, 6.14(F) – Supporting

1. Which of the following could be damaging to a person's credit history?

   I. Having 15 open accounts with balances close to their limits

   II. Making minimum monthly payments on the due date

   III. Having multiple hard inquiries on a credit report in a short period of time

   IV. Having multiple addresses listed in personal information

   Ⓐ I and III only

   Ⓑ II and IV only

   Ⓒ I, III, and IV only

   Ⓓ I, II, III, and IV

2. How many free credit reports is a person allowed in one year?

   Ⓕ One

   Ⓖ Two

   Ⓗ Unlimited

   Ⓙ None

3. Wendy wants to improve her credit score. Which of the following actions will help Wendy improve her score?

   Ⓐ Pay the minimum amount due on credit cards and loans.

   Ⓑ Close any accounts that have a zero balance.

   Ⓒ Make all monthly payments on or before the due date.

   Ⓓ Take out additional loans to pay off current debt.

4. Which of the following does NOT remain on a person's credit report for 7 years?

   Ⓕ Late payments

   Ⓖ A personal inquiry

   Ⓗ Bankruptcy

   Ⓙ Collection accounts

5. What is one advantage of using a credit card over a debit card to pay for a purchase?

   Ⓐ Using a credit card is never an advantage over using a debit card.

   Ⓑ Using a credit card allows the purchase to be paid over time so it is less expensive.

   Ⓒ Using a credit card is easier than using a debit card.

   Ⓓ Using a credit card responsibly builds credit history, which can positively impact credit worthiness.

6. Which of the following is NOT a reason to establish a positive credit history?

   Ⓕ Employers may look at a person's credit history before hiring.

   Ⓖ Colleges may look at a person's credit history before accepting them into the school.

   Ⓗ Apartment managers or landlords may look at a person's credit history before renting.

   Ⓙ Utility companies may look at a person's credit history before offering services.

Name _____

Standards 6.14(B) – Supporting, 6.14(D), 6.14(E) – Supporting, 6.14(F) – Supporting

Unit 41 Motivation Station

Sort the phrases below into two groups, *Positive Effect* and *Negative Effect*, depending on the effect the action has on a person's credit history.

a. Open new credit accounts

b. Pay off current account

c. Pay entire balance owed at the end of every month

d. Skip payments if cannot afford to pay

e. Close all accounts with a zero balance

f. Make payments on or before due date

g. Keep account balances at or below 25% of credit limits

h. File for bankruptcy

i. Unemployed for a long period of time

j. Have several years of responsible credit history

k. Have multiple hard inquiries for credit report

l. Dispute an unknown account and have it removed from credit report

m. Keep a steady job

n. Have one or more accounts listed in collections

Positive Effect	Negative Effect

Unit 41 Homework

Name _____

Standards 6.14(B) – Supporting, 6.14(D), 6.14(E) – Supporting, 6.14(F) – Supporting

For 1–5, determine if the statement is *true* or *false*. If false, rewrite the statement so that it is true.

1. Checking one's own credit report does not affect the credit score.
   _____
   _____

2. A credit score is the only thing that matters to a lender.
   _____
   _____

3. Credit scores are only used by lenders.
   _____
   _____

4. A person's credit score may change frequently.
   _____
   _____

5. Negative marks on a credit report can easily be removed.
   _____
   _____

6. Give three reasons why it is important to maintain a good credit history.
   _____
   _____
   _____

7. What is a delinquency? How does this affect your credit score?
   _____
   _____
   _____
   _____

8. Explain the differences between a debit card and a credit card.
   _____
   _____
   _____
   _____
   _____

9. Explain why information regarding a debit card is not part of a credit report.
   _____
   _____
   _____
   _____
   _____

10. List 5 things found in a credit report.
    _____
    _____
    _____
    _____
    _____

Connections

Explain how losing 10 points per day for a late assignment is like making a late payment on a credit card bill.

Name _____

Standards 6.14(G) – Supporting, 6.14(H) – Supporting

Unit 42 Vocabulary Activity

The following terms and phrases will be used throughout this unit.

annual salary	federal minimum wage	loans	scholarships
associate's degree	financial aid	master's degree	vocational training
bachelor's degree	grants	occupation	wages
doctorate	income	post-secondary education	work-study
Federal Application for Student Aid (FAFSA)			

Use the clues to unscramble the vocabulary terms below. Not all terms will be used.

1. m n i c e o — Any money earned by employment or investment _____

2. a n s r t g — Financial aid, typically based on need, not requiring repayment _____

3. c l a f i i a n n i a d — Money given or loaned to assist in paying for college tuition _____

4. l o s p a r h s c h s i — Money awarded based on academic or other achievement, not requiring repayment _____

5. a u l a n n a s r l a y — Compensation for work, expressed as a yearly sum _____

6. s c s e a i o t s a g e e e r d — Degree typically earned after 2 years of study from a community or career college _____

7. s a h c e o r b l e d g e e r — Degree typically earned after 4 years of study from a college or university _____

8. r k o w s u t y d — A student aid program that provides a part-time job while in school to help pay expenses _____

9. f s a a f — Free application used to determine eligibility for federal student aid _____

10. o t r c o e d a t — Highest post-graduate degree awarded by a university _____

11. e a f d l r e i m m u n i m g w a e — Lowest hourly amount allowed by federal government to be paid to a worker _____

12. s a l n o — Financial aid, typically based on need and credit worthiness, repaid with interest _____

13. a m r s e s t e g d r e e — Post-graduate degree typically earned after 3 years of study from a college or university _____

14. o a v t c n o a i l g t i n r a i n — Training provided for a specific field or occupation _____

©2014 mentoringminds.com   motivation**math**™ LEVEL 6   ILLEGAL TO COPY   333

Unit 42 Practice

Name _____

Standards 6.14(G) – Supporting, 6.14(H) – Supporting

Hillary and Matt are sixth-grade students at Tejas Middle School. The class is studying occupations and the levels of education needed to succeed in the future job market, as well as how to pay for their education. The teacher gives the students a survey to complete about their future plans. Hillary's and Matt's completed surveys are shown below.

Student Career Interest Survey Name __Hillary__

1. What are your favorite subjects in school? If you don't have a favorite subject, what is your favorite activity outside of school?

   I like math and science and music.

2. What areas or fields of careers interest you right now? List at least 2, if possible.

   Health care or teaching

3. What would you like your work environment to be like? For example, do you prefer being indoors or outdoors? Do you like to work alone or would you prefer working with a group?

   I like working with other people and helping people. I like being outside, but I don't think I want to

   work outside. I want to work somewhere that is always fast and I'm always busy so I don't get bored.

Student Career Interest Survey Name __Matt__

1. What are your favorite subjects in school? If you don't have a favorite subject, what is your favorite activity outside of school?

   I like my shop class. We get to build stuff and that's fun.

2. What areas or fields of careers interest you right now? List at least 2, if possible.

   I don't really know anything. I want to be a builder or something.

3. What would you like your work environment to be like? For example, do you prefer being indoors or outdoors? Do you like to work alone or would you prefer working with a group?

   Sometimes I like working with people but most times I just want to work by myself. I like to be

   outside a lot and work on stuff that's broken to see if I can fix it. I want to work somewhere that I can

   just do stuff and nobody tells me what to do.

---

1 Visit the *CareerOneStop* website, www.careerinfonet.org, and use the instructions below to complete the following table.

- Click on the link *Occupation Information*.
- Click on the first bullet point, *Occupation Profile*.
- In the *Keyword Search* box, enter the occupations listed in the table on the next page, one at a time, and search.
- Highlight the occupation you entered and click *Continue*.
- Scroll to find *Texas* and click *Continue*.
- Use the information on the page to complete the table.
- Click the link *Occupation Profile* on the left-hand side of the page to return to the search box.

Name _____

Standards 6.14(G) – Supporting, 6.14(H) – Supporting

Unit 42 Practice

Hillary's Career Interests

Occupation	Texas median annual salary ($)	Education/Training
Licensed practical or vocational nurse		
Registered nurse		
Nurse practitioner		

Matt's Career Interests

Occupation	Texas median annual salary ($)	Education/Training
Cabinetmaker and bench carpenter		
Carpenter		
Welder, cutter, solderer, brazer		

Hillary wants to become a registered nurse. She will attend a four-year university, and the expected cost to attend for four years is $66,740 plus books and supplies.

2. Visit the *Federal Student Aid* website of the U.S. Department of Education at www.college.gov/types. Begin by watching the 2-minute video about student financial aid. Then answer these questions.

   a. List the 3 types of federal student aid.

   _____

   b. What is the first step Hillary must take to receive federal student aid?

   _____

   c. Click on the link *Grants* under *Aid and Other Resources from the Federal Government*. List the 4 types of federal grants. Read more about each grant by clicking on the link.

   _____

   _____

   d. Which 2 types of grants should Hillary apply to receive?

   _____

3. Matt plans to begin working as a welder at age 20 and retire at 70. Hillary plans to begin working as a registered nurse at age 22 and retire at 70. What will be the difference in Matt's and Hillary's lifetime earnings? Show all work.

Unit 42 Assessment

Standards 6.14(G) – Supporting, 6.14(H) – Supporting

Use the information to answer questions 1–3.

Elena is graduating from high school. She will attend a private, 4-year university and study to become a Certified Public Accountant. The annual cost for her tuition, room and board, and books is $22,255. Elena's parents have a college fund with $54,000 to help with the costs. Elena received a scholarship for $500 per semester for four years.

1 What is Elena's annual cost after her parents' contribution and her scholarship?

Ⓐ $31,020

Ⓑ $8,255

Ⓒ $7,755

Ⓓ $8,755

2 What are Elena's options for paying the remainder of her college tuition each year?

Ⓕ Grants and student loans only

Ⓖ Grants, additional scholarships, and loans only

Ⓗ Student or parent loans only

Ⓙ Grants and additional scholarships only

3 Which method of paying the remainder of her college tuition each year will cost Elena the least?

Ⓐ Loans, because she does not have to begin repayment until 6 months after graduation

Ⓑ Scholarships, because there are no eligibility requirements

Ⓒ Work-study program, because she earns her own money by working on campus

Ⓓ Grants, because she is not required to repay the money given to her

Use the information to answer questions 4–6.

Sabino dropped out of school when he was 17 and started working for a construction company. He earns $11.89 per hour working 40 hours per week. Last year he earned $24,255. Now he wants to earn his GED so he can go to college and become a construction manager, which requires a bachelor's degree. His potential job earnings are $72,500.

4 Which statement best describes Sabino's potential earnings with a bachelor's degree?

Ⓕ He could potentially earn over $48,000 more annually with a bachelor's degree.

Ⓖ He will earn less per hour with a bachelor's degree because he will work more hours.

Ⓗ His monthly income with a bachelor's degree will be less than twice his current monthly income without a diploma.

Ⓙ His lifetime income will more than quadruple with a bachelor's degree compared to his lifetime income without a diploma.

5 Sabino decides to continue working and attend night school. If he completes his degree and begins working as a construction manager at age 30, what will his lifetime earnings be? Include the time he spends working before he becomes a manager and assume he will retire at age 65.

Ⓐ $2,537,500   Ⓒ $2,683,030

Ⓑ $2,852,815   Ⓓ $3,480,000

6 If Sabino does not return to school and continues working as a construction laborer until age 65, which of the following is true?

Ⓕ His lifetime earnings will be more because he will work longer at one job.

Ⓖ His lifetime earnings as a laborer will be more than 50% of his earnings as a manager.

Ⓗ His lifetime earnings as a laborer will be about 40% of his earnings as a manager.

Ⓙ His lifetime earnings will be no different if he does not return to school.

Name _____

Standards 6.14(G) – Supporting, 6.14(H) – Supporting

Unit 42 Motivation Station

Will Work For...

Visit the website *Texas Reality Check* at www.texasrealitycheck.com to complete this page. Follow the instructions and record your selections for each page. Hover over or click on each section on a page to learn more information about monthly costs. Use the *Next* and *Back* buttons to navigate between the pages.

1. Begin by clicking on 1: Reality Check. Select a city you would like to live in or near.
   _____

2. Housing: Select the type of housing you would like. _____

3. Utilities: Determine which utilities you will need/want. Electricity, water, and gas are pre-selected and may not be removed.
   _____

4. Food: Choose a food option that best fits your needs. _____

5. Transportation: Determine how you will get to work and other places by choosing transportation. If you have chosen to live in a smaller community, a public bus may not be available to you. _____

6. Clothing: Select the type(s) of clothing allowance you would like each month.
   _____

7. Health care: Choose how you will address health care. This is to address your budget. It does not necessarily mean you will go to the doctor every month.
   _____

8. Personal: Will you have a gym membership? How much do you spend each month on hair styling and beauty products? Choose one or more categories to meet your personal needs.
   _____

9. Entertainment: Select one or more categories for an entertainment allowance.
   _____

10. Miscellaneous: This page allows you to make choices for your life such as saving for vacations, keeping a pet, etc. _____

11. Savings: Savings is a very important, and often overlooked, part of a budget. Choose the savings amount that makes you most comfortable. _____

12. Student Loan Debt: This page is optional, if you believe you will need student loans to pay for college. Select your degree type and institution type to budget a payment for student loan debt. _____

13. Total Expenses: What are your total monthly expenses? What is the annual salary you need to cover your expenses and taxes? _____
    _____

Continue to *Find Careers* on the right side of the page. Explore the types of jobs available for your education choice and that will earn enough money to pay your total expenses.

Unit 42 Homework

Standards 6.14(G) – Supporting, 6.14(H) – Supporting

Use the table to answer questions 1 and 2.

Median Annual Salaries Based on Education Required

Occupation	Education Required	Median Annual Salary ($)	Education Costs
Restaurant cook	Less than high school	21,774	none
Accounting clerk	High school/GED; on-the-job training or associate's degree	29,991	less than $25,000
Elementary school teacher	Bachelor's degree	50,110	less than $70,000
Physical therapist	Master's degree	72,790	less than $105,000
Astronomer/astrophysicist	Doctorate	81,208	less than $150,000

1. For each occupation shown in the table above, calculate the lifetime earnings for a person in that position, based on the criteria given, and the number of years the person works in the position.

   a. Restaurant cook – begins working at age 17 and retires at age 68

   b. Accounting clerk – begins working at age 18 and retires at age 65

   c. Elementary school teacher – begins working at age 22 and retires at age 55

   d. Physical therapist – begins working at age 24 and retires at age 60

   e. Astronomer/astrophysicist – begins working at age 30 and retires at age 62

2. Which occupation would you choose? Why? Consider all factors in your response, not just salary.

_____

_____

Connections

Conduct an interview with someone you know who attended college. Ask questions such as, "How did you know what you wanted to study in college?" "How did you pay for college?" "How long did it take to earn your degree?" Summarize the interview in a written or oral report to be presented to the class.

Performance Assessment A: Weather or Not, It's Mathematics!

Performance Assessment B: Cardinal School Store

Name _____

Weather or Not, It's Mathematics!

Performance Assessment A

Problem Stimulus

Madeline sat down for a typical day in her sixth-grade mathematics class at Westside Middle School in Mountain Town. Mrs. Baldwin was about to start her lesson for the day.

Madeline turned to her friend Kevin and said, "I never understand the weather. I feel like sometimes here in Mountain Town it will be snowing one day and then warm and sunny the next! Some months it rains for days, and some months it's always dry. I never know what to wear."

Kevin said, "I agree. I heard that we're going to learn more about weather in our science class. Maybe we will learn how to figure out what the weather will be."

Mrs. Baldwin overheard the students talking. "Madeline and Kevin, I hear you talking. I know Mr. Swanson, your science teacher, is going to teach you about weather. You are going to learn all the different types of weather and find causes," said Mrs. Baldwin. "Do you think there is any mathematics involved in weather?"

"I don't think so, Mrs. Baldwin," said Kevin, "weather takes place in nature, which is science."

"Well, let's take a look. I can show you many different things that you've studied this year that are used to predict the weather. I think you will be surprised to discover how much mathematics relates to the weather," said Mrs. Baldwin.

"Let's see, Mrs. Baldwin! I don't know if I believe you, but I'd sure like to know what to wear!" exclaimed Madeline.

Task Overview

The sixth grade will soon study weather in science class. Mrs. Baldwin is on a mission to show her students that mathematics is useful in weather forecasting.

Performance Assessment A

Weather or Not, It's Mathematics!

Performance Task

Part A

"Well," said Mrs. Baldwin, "one of the most important things that you will study in your weather unit is temperature. Here in Mountain Town, we have a broad range of temperatures throughout the year. Here is a table that shows the lowest temperature measured each month over the last year, in degrees Fahrenheit."

Monthly Low Temperatures (°F)

Jan.	Feb.	March	April	May	June	July	Aug.	Sept.	Oct.	Nov.	Dec.
-14	-9.1	8	29	56.5	58	56	45	30	14	-9	-8.5

1. Using a number line, order the low temperatures from lowest to highest.

2. Find the difference between the coldest temperature and the warmest temperature in the table. Use words or numbers to explain how you got your answer.

3. The quotient of the lowest recorded temperature in November and -3 yields the second lowest recorded temperature of the month. What was the second lowest recorded temperature for November? Show all work.

4. The highest recorded temperature in October is 2.5 times the lowest recorded temperature in April. What is the highest temperature in October? Show all work.

Name _____

Weather or Not, It's Mathematics!

Performance Assessment A

Part B

"I can see what you're saying about temperatures; that is all mathematics," said Madeline, "but what about rain and snow? How could that be mathematics?"

Mrs. Baldwin began writing problems on the board. "Remember all those problems we did earlier this year with fractions, decimals, and percents? Take a look."

1. During the month of April in Mountain Town, it rained $\frac{1}{3}$ of the days. In June, it rained 0.15 of the days. Lastly, in November, it rained 20 percent of the days. Order the months from the one with the lowest number of days of rain to the one with the highest number of days of rain. Explain how you got your answer.

2. The National Weather Service collected data for the month of April. In Mountain Town, it rained 7.5 inches in the month of April. What is the average rate of rainfall per day for the month? Explain how you got your answer.

3. In November, Mountain Town received snow on 30% of the days in the month. On how many days did it snow in November? Show all work.

4. Over the course of the last year, which was not a leap year, there were 73 days on which measurable precipitation fell in Mountain Town. Express the part of the year with measurable precipitation as a fraction, a decimal, and a percent. Express answers in lowest terms.

5. The record rate of snowfall in Mountain Town is 2.5 inches per hour. What is this rate in feet per day? Show all work.

Performance Assessment A

Name _____

Weather or Not, It's Mathematics!

Part C

"You know, Mrs. Baldwin, you're right. This weather stuff is all mathematics. I think we could even use graphing and equations to display data from storms," said Kevin.

"You're right, Kevin. Let's look at some data from snowstorms," said Mrs. Baldwin.

During one particular snowstorm, the weather team at the Mountain Town news channel measured the depth of the snow over time. The snow fell at a constant rate. Below is the depth of snow the weather team measured during the storm.

Number of Hours, h	Amount of Snow, s (in.)
0	0.0
1	0.5
2	1.0
3	1.5
4	2.0

1 Is the relationship between the number of hours it snowed and the snow accumulation additive or multiplicative? Explain.

2 Identify the independent and dependent quantities from the table.

Write an equation that can be used to find the depth of the snow at the news studio given the number of hours it snowed.

Name _____

Weather or Not, It's Mathematics!

Performance Assessment A

3 Represent the situation from the table and equation on the graph provided.

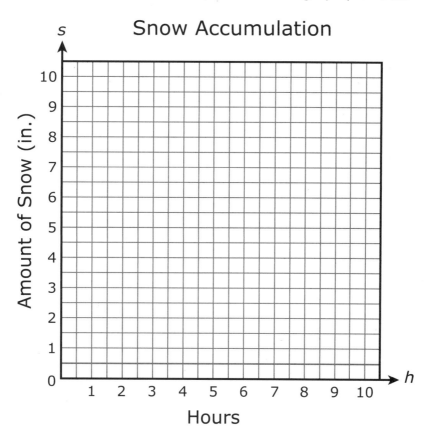

4 Use the graph to answer the following.

a. How much snow, in inches, accumulated in 6 hours?

b. If $4\frac{1}{2}$ inches of snow accumulated, how many hours had it been snowing?

c. At the same rate, how long would it take for one foot of snow to accumulate? Explain how you determined your answer.

Name: _____

Performance Assessment A: Weather or Not, It's Mathematics! Scoring Rubric

Task	Score Point: 1 Undeveloped	Score Point: 2 Developing	Score Point: 3 Developed	Student's Score (Circle)
A.1 Order temperatures on a number line.	Student places 0–5 values correctly on a number line.	Student places 6–10 values correctly on a number line.	Student places 11–12 values correctly on a number line.	Score: 1 2 3
A.2 Find the difference between positive and negative temperatures.	Student does not record the right answer and does not provide the correct explanation.	Student records the correct answer, but cannot explain the work.	Student records the correct answer and correctly explains the work.	Score: 1 2 3
A.3 Divide a negative temperature by a negative number.	Student does not divide.	Student gets -3 for an answer, which does not reflect the correct sign.	Student records the correct answer and shows correct work.	Score: 1 2 3
A.4 Multiply a positive temperature by a positive rational number.	Student does not show understanding of multiplication.	Student multiplies 29 and 2.5, but does not get the correct answer.	Student records the correct answer and shows correct work.	Score: 1 2 3
B.1 Order fractions, decimals, and percents of precipitation.	Student incorrectly orders the values and cannot explain the work.	Student correctly orders the values, but does not correctly explain the work. OR Student incorrectly orders the values, but has the correct explanation of the work.	Student records the correct answer and can explain the work.	Score: 1 2 3
B.2 Find the unit rate of precipitation.	Student does not know how to find the unit rate.	Student attempts to find the unit rate correctly, but does not get the correct answer. OR Student knows to divide, but reverses the dividend and divisor.	Student records the correct answer and shows correct work.	Score: 1 2 3
B.3 Find the number of days that it snowed, given the percent of snow and the number of days in the month.	Student does not set up the problem correctly and does not get the correct answer.	Student sets up the problem correctly, but does not get the correct answer. OR Student does not have the correct work, but does have 9 days as the answer.	Student records the correct answer and shows correct work.	Score: 1 2 3

Name: _____

Performance Assessment A: Weather or Not, It's Mathematics! Scoring Rubric

Task	Score Point: 1 Undeveloped	Score Point: 2 Developing	Score Point: 3 Developed	Student's Score (Circle)
B.4 Find equivalent fractions, decimals, and percents for the days of precipitation over the course of a year.	Student shows how to make 0 or 1 fraction, decimal, or percent from the given information.	Student shows how to make 2 of the forms (fraction, decimal, or percent) from the given information.	Student records the correct answers.	Score: 1 2 3
B.5 Convert inches per hour to feet per day.	Student shows no understanding of how to convert units of measure.	Student is partially able to convert the unit. Student is able to find the inches per day or the feet per hour.	Student records the correct answer and shows correct work.	Score: 1 2 3
C.1 Identify the relationship in the table.	Student identifies the relationship as additive.	Student identifies the relationship as multiplicative, but is unable to justify the answer.	Student identifies the relationship as multiplicative and provides the correct explanation.	Score: 1 2 3
C.2 Identify independent and dependent quantities, and write an equation from the table.	Student is unable to identify independent and dependent quantities or record an equation.	Student correctly identifies independent and dependent quantities, but incorrectly records the equation, or vice versa.	Student correctly identifies independent and dependent quantities, and records the correct equation.	Score: 1 2 3
C.3 Graph an equation for hourly snowfall.	Student is unable to correctly draw the graph.	Student incorrectly plots one or more points from the table.	Student correctly graphs the line showing continuous values.	Score: 1 2 3
C.4 Answer questions using a graph.	Student answers 0 or 1 question correctly.	Student answers 2 questions correctly.	Student answers all three questions correctly.	Score: 1 2 3
			Student's Total Score	

35–39 points = Proficient   29–34 points = Satisfactory   23–28 points = Below Standard   0–22 points = Unsatisfactory

Name _____

Cardinal School Store

Performance Assessment B

Problem Stimulus

The student council at Downtown Middle School was beginning their monthly meeting. "Welcome!" said Mr. Smith, the student council sponsor. "Today we have the exciting task of talking about money!"

"That's right," chimed in Michael, the student council president. "It's that time of year when we need to start raising funds for upcoming school events. We need to think of a good way to fill up the student bank account."

"I have an idea," said Michelle. "I've been working at the school store, and I've had many students ask for school spirit gear. They want to represent the Cardinals, wear and use things in our school colors, and show that they are proud to be DMS students!"

"I think that's a great idea!" exclaimed Michael. "We sell a lot at the school store. Kids are always buying snacks, pencils, notebooks, and drinks. I think Cardinal spirit gear would be a great seller!"

"As the treasurer," said Shonda, "I really like this idea, but we've got to do our research. I know these things can get expensive. We should compare prices from different companies before making any decisions."

"I like where this conversation is going, and I love supporting the Cardinals. Let's start researching. What should we sell? How much will it cost us? How should we price the items?" questioned Michael.

"That's the spirit — school spirit that is!" laughed Mr. Smith. "Let's find answers to our questions so we can spread Cardinal spirit!"

Task Overview

The student council is excited to launch the sale of school spirit gear in the school store. First, they need to research and determine which items they will sell. Then, they need to make comparisons among different companies and determine the cost of the items. Finally, the student council needs to determine the profit from their sales.

Performance Assessment B

Cardinal School Store

Performance Task

Part A

1. The student council surveys students before they begin ordering for the school store. They research five popular spirit items, and then ask 160 sixth-grade students which item they would most likely purchase. The results of the survey are shown below. Complete the relative frequency table, rounding values to the nearest hundredth.

Spirit Items Most Likely to Purchase

Item	Number of Students (Frequency)	Relative Frequency
Pencil	40	
Sticker	52	
Sweatshirt	2	
T-shirt	40	
Temporary tattoo	26	

What is the mode of the data shown in the relative frequency table? Explain how you determined your answer.

2. Create a percent bar graph using the data in the relative frequency table. Include a key to accompany your graph.

The percent of students selecting *temporary tattoo* as the item they would purchase is approximately half the percent of students selecting which item?

Name _____

Cardinal School Store

Performance Assessment B

3 In the same survey, students were asked how many pencils they use each month.

Pencils Used per Month

Number of Pencils	Number of Students
0–2	20
3–5	30
6–8	65
9–11	40
12–14	10

Use the data from the table to create a histogram.

Which best describes the shape of the data, symmetric or skewed? Justify your answer.

4 Finally, 15 students were randomly selected and asked how much they would pay for a Cardinal spirit T-shirt. The results are shown in the stem-and-leaf plot.

Price of a T-shirt (in dollars)

```
Stem | Leaves
  0  | 5 6 7 7
  1  | 0 0 1 1 2 6 7 9 9
  2  | 0 0
```

0 | 5   represents 5

What is the median price shown in the plot? Explain how you determined your answer.

What is the spread of the data distribution shown in the plot?

Performance Assessment B

Cardinal School Store

Part B

Based on their initial research, the student council decides to sell Cardinal stickers, pencils, and T-shirts in the school store. Their next step is to determine the cost of the items.

1. Three companies sell stickers with the Cardinal logo printed in school colors. The table below shows the number of stickers and the cost for each company.

Sticker Prices

Company	Number of Stickers	Cost ($)
Sticks-a-Lot	200	50
Mr. Stickems	100	28
Sticky's	300	72

Which company offers the best deal? Show all work.

2. The student council determined that the best price for pencils was with Get-to-the-Point Pencil Company. The student council orders 15 packages of one dozen pencils for $18. What is the cost per pencil? Show all work.

3. The total cost for a clothing order from Spirit USA Clothes is $1,450. The company offers a discount to reduce the total to $1,160, a savings of $290. What percent of the total cost is the student council saving on the clothing order? Show all work.

Name _____

Cardinal School Store

Performance Assessment B

Part C

The student council places their first order and prepares to sell spirit items in the school store. They need to determine the sale price of each item to be sold to ensure they make a profit.

1. The student council spends $192 on 800 Cardinal stickers. The student council wants to make a profit of $200 selling the stickers. What is the total amount of sales needed to make a profit of $200?

   Based on your response above, write and solve an inequality that can be used to determine the prices, p, the student council can charge per sticker to make a profit of at least $200.

2. The student council spent $1,160 on T-shirts. They sell the T-shirts for $10 each. Write an equation that can be used to determine t, the number of T-shirts the student council needs to sell to pay the cost.

   Graph the solution set to the equation written above on a number line.

3. The student council wants to make more than $36 in sales on pencils. The council plans to sell the pencils for $0.20 each. Write an inequality that can be used to determine p, the numbers of pencils that can be sold to make more than $36 in sales.

   Which of the following values for p are true for this scenario? Justify your responses.

   a. 180

   b. 200

   c. 210.5

   d. 1.8

   e. 250

Name: _____

Performance Assessment B: Cardinal School Store Scoring Rubric

Task	Score Point: 1 Undeveloped	Score Point: 2 Developing	Score Point: 3 Developed	Student's Score (Circle)
A.1 Create a relative frequency table and identify the mode of a data set.	Student incorrectly completes the table and is unable to identify the mode of the data.	Student correctly identifies the mode of the data, but incorrectly completes the table, or vice versa.	Student correctly completes the table and identifies the mode of the data.	Score: 1 2 3
A.2 Create a percent bar graph and use the graph to answer a question.	Student is unable to create the percent bar graph, and incorrectly answers the question.	Student correctly creates the percent bar graph, but incorrectly answers the question, or vice versa.	Student correctly creates the percent bar graph and answers the question.	Score: 1 2 3
A.3 Represent data from the survey in a histogram and describe the shape of the data distribution.	Student is unable to represent data correctly in a histogram.	Student represents data correctly in a histogram, but is unable to describe and justify the shape of the data distribution.	Student represents data correctly in a histogram, and is able to describe and justify the shape of the data distribution.	Score: 1 2 3
A.4 Use a stem-and-leaf plot to find the median and spread of the data distribution.	Student is unable to find the median or spread of the data distribution.	Student correctly finds the median of the data distribution, but incorrectly finds the spread, or vice versa.	Student correctly finds the median and spread of the data distribution.	Score: 1 2 3
B.1 Find the unit rate and determine the best deal.	Student is unable to correctly determine the unit rates.	Student determines some, but not all, of the unit rates correctly, or is unable to determine the best deal.	Student determines the unit rates and the best deal correctly.	Score: 1 2 3
B.2 Find the unit rate.	Student is unable to correctly set up the computation to determine the unit rate.	Student correctly sets up the computation to determine the unit rate, but makes a computation error.	Student correctly determines the unit rate.	Score: 1 2 3
B.3 Find the percent when given the part and the whole.	Student is unable to correctly set up the computation to determine the percent.	Student correctly sets up the computation to determine the percent, but makes a computation error.	Student correctly determines the percent.	Score: 1 2 3

Name: _____

Performance Assessment B: Cardinal School Store Scoring Rubric

Task	Score Point: 1 Undeveloped	Score Point: 2 Developing	Score Point: 3 Developed	Student's Score (Circle)
C.1 Write and solve an inequality.	Student is unable to use the given information to form an inequality.	Student forms only part of the inequality correctly, resulting in an incorrect solution.	Student correctly forms and solves an inequality.	Score: 1 2 3
C.2 Write an equation and graph the solution.	Student is unable to use the given information to form an equation.	Student forms only part of the equation correctly, resulting in an incorrect graph of the solution.	Student correctly forms the equation and graphs the solution.	Score: 1 2 3
C.3 Write an inequality and determine whether given values make the inequality true.	Student is unable to use the given information to form an inequality.	Student forms only part of the inequality correctly, and is therefore unable to correctly determine whether given values make the inequality true.	Student correctly forms the inequality and is able to correctly determine whether given values make the inequality true, justifying the responses.	Score: 1 2 3
			Student's Total Score	

27–30 points = Proficient   22–26 points = Satisfactory   18–21 points = Below Standard   0–17 points = Unsatisfactory

Chart Your Success

Place a check (✓) in the box if your answer is correct.

Unit / Page	Objective	1	2	3	4	5	6	Total Correct	Total Possible
Unit 1 Page 10	Classify numbers 6.2(A)								4
Unit 2 Page 18	Identify a number, its opposite, and its absolute value 6.2(B)								6
Unit 3 Page 26	Locate, compare, and order numbers using a number line 6.2(C)								5
Unit 4 Page 34	Order rational numbers 6.2(D)								5
Unit 5 Page 42	Interpret fractions as division 6.2(E)								6
Unit 6 Page 50	Dividing by a rational number and multiplying by its reciprocal 6.3(A)								6
Unit 7 Page 58	Determine results when multiplying by a fraction 6.3(B)								6
Unit 8 Page 66	Represent integer operations with concrete models 6.3(C)								4
Unit 9 Page 74	Add, subtract, multiply, and divide integers 6.3(D)								6
Unit 10 Page 82	Multiply and divide positive rational numbers 6.3(E)								5
Unit 11 Page 90	Additive and multiplicative relationships 6.4(A)								4
Unit 12 Page 98	Solve problems involving ratios and rates 6.4(B)								5
Unit 13 Page 106	Ratios as multiplicative comparisons of two quantities 6.4(C)								5

©2014 mentoringminds.com   motivationmath™ LEVEL 6   ILLEGAL TO COPY

Chart Your Success

Place a check (✓) in the box if your answer is correct.

Unit / Page	Description	1	2	3	4	5	6	7	Total Correct	Total Possible
Unit 14 Page 114	Rates as the comparison by division of two quantities 6.4(D)									5
Unit 15 Page 122	Represent ratios, percents, and benchmark fractions 6.4(E), 6.4(F)									5
Unit 16 Page 130	Generate equivalent fractions, decimals, and percents 6.4(G)									7
Unit 17 Page 138	Convert units within a measurement system 6.4(H)									6
Unit 18 Page 146	Represent problems involving ratios and rates 6.5(A)									6
Unit 19 Page 154	Solve real-world problems involving percents 6.5(B)									6
Unit 20 Page 162	Use equivalent fractions, decimals, and percents to show equal parts 6.5(C)									6
Unit 21 Page 170	Identify and write equations for independent and dependent quantities 6.6(A), 6.6(B)									5
Unit 22 Page 178	Represent a situation in the form of $y = kx$ or $y = x + b$ 6.6(C)									4
Unit 23 Page 186	Generate equivalent numerical expressions using order of operations 6.7(A)									5
Unit 24 Page 194	Expression and equations 6.7(B), 6.7(C)									5
Unit 25 Page 202	Generate equivalent expressions using the properties of operations 6.7(D)									5
Unit 26 Page 210	Extend previous knowledge of triangles and their properties 6.8(A)									4

Chart Your Success

Place a check (✓) in the box if your answer is correct.

Unit / Page	Objective	1	2	3	4	5	6	Total Correct	Total Possible
Unit 27 Page 218	Model area formulas by decomposing and rearranging parts 6.8(B)	1	2	3	4			Total Correct	4
Unit 28 Page 226	Write equations and solve problems involving area and volume 6.8(C), 6.8(D)	1	2	3	4	5		Total Correct	5
Unit 29 Page 234	Write equations and inequalities and represent their solutions 6.9(A), 6.9(B)	1	2	3	4	5	6	Total Correct	6
Unit 30 Page 242	Write real-world problems given equations or inequalities 6.9(C)	1	2	3	4			Total Correct	4
Unit 31 Page 250	Model and solve equations and inequalities 6.10(A)	1	2	3	4	5		Total Correct	5
Unit 32 Page 258	Determine if given values make equations or inequalities true 6.10(B)	1	2	3	4	5	6	Total Correct	6
Unit 33 Page 266	Graph points in all four quadrants 6.11(A)	1	2	3	4	5	6	Total Correct	6
Unit 34 Page 274	Represent numeric data graphically 6.12(A)	1	2					Total Correct	2
Unit 35 Page 282	Describe the center, spread, and shape of data distribution 6.12(B)	1	2	3	4			Total Correct	4
Unit 36 Page 290	Summarize numerical data with numerical summaries 6.12(C)	1	2	3	4	5	6	Total Correct	6
Unit 37 Page 298	Summarize categorical data with numerical and graphical summaries 6.12(D)	1	2	3				Total Correct	3
Unit 38 Page 306	Interpret summaries of numeric data 6.13(A)	1	2	3	4			Total Correct	4
Unit 39 Page 314	Distinguish between situations that yield data with and without variability 6.13(B)	1	2	3	4			Total Correct	4

Chart Your Success

Place a check (✓) in the box if your answer is correct.

									Total Correct	Total Possible
Unit 40 Page **324**	Compare features of a checking account; balance a check register 6.14(A), 6.14(C)	1	2	3						**3**
Unit 41 Page **330**	Debit cards vs. credit cards; credit history; credit reports 6.14(B), 6.14(D), 6.14(E), 6.14(F)	1	2	3	4	5	6			**6**
Unit 42 Page **336**	Paying for college; comparing occupations and lifetime earnings 6.14(G), 6.14(H)	1	2	3	4	5	6			**6**

Math Glossary

A

absolute value – the distance of a number from zero on a number line

acute angle – an angle measuring less than 90°

acute triangle – a triangle with three acute angles

addend – numbers that are added

additive relationship – a relationship among a set of ordered pairs in which the same value is added to each *x*-value to find the corresponding *y*-value

The table shows an additive relationship because 4 is added to each *x*-value to find each *y*-value.

x	0	1	2	3
y	4	5	6	7

adjacent – side-by-side; adjoining

algebraic equation – a number sentence that contains at least one variable and uses the equal sign to show that two expressions are equivalent; for example, $2a = 7$

algebraic expression – an expression that contains at least one variable; for example, $a + b$ or $5n$.

algorithm – a step-by-step process for solving a problem

altitude – the perpendicular distance from the base of a shape to the highest point of the shape

angle – a figure formed by 2 rays that share a common endpoint

annual fee – a yearly fee charged by a credit card company for the use of a credit card

annual percentage rate (APR) – the yearly cost of a financed amount, including interest and fees, expressed as a percentage rate

annual salary – compensation for work, expressed as a yearly sum

approximate – *verb*: to estimate; to come close to; *adjective*: almost exact or correct

area – the number of square units needed to cover a surface

area of base (B) – in a three-dimensional figure, the area, in square units, of the base of the figure

ascend – to increase

associate's degree – degree typically earned after two years of study from a community or career college

associative property – a property of addition or multiplication in which the grouping of the addends or factors does not change the outcome of the operation

$$(a + b) + c = a + (b + c)$$
$$(1 + 2) + 3 = 1 + (2 + 3)$$
or
$$(a \cdot b) \cdot c = a \cdot (b \cdot c)$$
$$(1 \cdot 2) \cdot 3 = 1 \cdot (2 \cdot 3)$$

ATM – acronym for automated teller machine; an electronic banking device that allows customers to complete financial transactions without the aid of a teller

attribute – a characteristic or property of a shape or thing

axis/axes – the horizontal (*x*-axis) and vertical (*y*-axis) number lines on a coordinate plane

B

bachelor's degree – a degree typically earned after four years of study from a college or university

balance (of a bank account) – the total amount of money in a bank account at any given time

base (2-D figure) – the side of a polygon from which an altitude is drawn

base (3-D figure) – the face or faces of a geometric figure from which an altitude is drawn

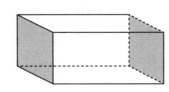 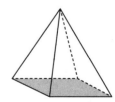

base (of an exponent) – factor appearing the number of times shown by an exponent; for example, 9 is the base in the expression 9^3

base-ten place value system – a numbering system based on 10 in which the value of each place is 10 times as much as the place to its right and $\frac{1}{10}$ the value of the place to its left

benchmark fraction – a common fraction used to judge the approximate size of another fraction; benchmark fractions may include $\frac{1}{4}$, $\frac{1}{2}$, and $\frac{3}{4}$

borrow – to obtain or use money from a bank or other person in order to purchase goods or services

borrower – a person who borrows money

box plot – a graph that uses a box to represent the middle 50% of a data set and line segments at each end to represent the remainder of the data

braces – symbols used to indicate a set or subset of data; { }

brackets – symbols used to group part of a mathematical expression or equation; [ ]

C

capacity – a measure of the amount of liquid a container will hold

categorical data – observations that can be sorted into groups or categories, such as shoe colors or favorite vegetables

central tendency – any of the three measures (mean, median, mode) that represent the center of a set of data

check – a slip of paper that is filled out to allow a bank to take money from a checking account to pay for a purchase

check register – a table for recording information about transactions written on a bank account

checking account – an account at a financial institution that allows a person to deposit or withdraw money and write checks to pay for goods and services

coefficient – a number multiplied by a variable in an algebraic expression; for example, in the expression $6x$, 6 is a coefficient

collection account – a delinquent or past-due account that has been transferred to a collection agency

common denominator – a denominator that is the same in two or more fractions

commutative property – a property of addition or multiplication in which the sum or product stays the same when the order of the addends or factors is changed

$$a + b = b + a$$
$$7 + 8 = 8 + 7$$
or
$$a \cdot b = b \cdot a$$
$$3 \cdot 4 = 4 \cdot 3$$

compare – to determine whether two or more numbers or quantities are greater than, less than, or equal to one another

complementary angles – two angles whose measures sum to 90°

composite number – a whole number that has more than two whole-number factors; the number 10 is a composite number because it has more than two factors: 1, 2, 5, and 10

computation – a calculation of numbers

congruent – having the same size and shape

constant – a number that does not vary; for example, in the expression $x + 5$, 5 is a constant

continuous data – any data that has infinite values with connected data points, such as measurements

conversion – to change from one unit of measure to another

coordinate plane – a plane formed by two perpendicular number lines intersecting at 0

coordinates – an ordered pair of numbers (x, y) used to locate a point on a coordinate plane

credit – an amount of money that a lender or business allows a person to use to purchase goods and services with a promise to repay the money, usually with interest

credit card – a card, issued by a bank, store, or other business, that is used to borrow money or buy goods and services on credit; borrowers make regular payments to repay the amount of money borrowed with interest and fees

credit limit – the maximum amount a credit card will allow someone to borrow on a single card

credit report – an official record of the borrower's credit history; used by lenders to determine a borrower's credit worthiness

credit score – a numerical summary about the information contained in a credit report

credit union – a financial institution that is owned by its members; a credit union may offer options for checking accounts, savings accounts, and loans to its members

cube – a three-dimensional figure with six congruent square faces

cubed – raised to a power of 3; for example, 4 cubed = 4^3

cubic unit – a unit, shaped like a cube with dimensions of 1 unit × 1 unit × 1 unit, used to measure volume

customary system of measurement – the measurement system used most often in the United States

D

data – a collection of facts or information gathered by observing, questioning, or measuring, usually displayed on a chart, table, or graph

data point – a single fact or piece of information

debit – a payment made that removes money from a bank account

debit card – a card issued by a bank that a customer uses to pay for purchases with money directly from a checking account

debt – money, goods, or services owed to others

decimal number – a number that uses base-ten place value and a decimal point to show tenths, hundredths, thousandths, etc.

decompose – to break down or break apart into smaller parts

degree (°) – a unit of measure for angles and temperature

delinquency – a debt or other account for which a payment is past due

denominator – the bottom number in a fraction; the total number of equal parts

dependent quantity – a quantity whose value is determined by the value of the related independent quantity

x	1	2	3	4
y	8	9	10	11

dependent quantities: {8, 9, 10, 11}

deposit – money placed into a checking or savings account at a bank

descend – to decrease

difference – the answer to a subtraction problem

digit – one of the symbols 0, 1, 2, 3, 4, 5, 6, 7, 8, and 9 used to write numbers

dimensions – the measures of sides in a geometric figure

direct deposit – an electronic transfer of a payment directly from the payer's account to the recipient's account

discrete data – a set of finite data values with unconnected data points, such as a count or scores

distributive property – the property which states that multiplying a sum by a number is the same as multiplying each addend by the number and then adding the products

$$a(b + c) = ab + ac$$
$$3(4 + 2) = (3 \times 4) + (3 \times 2)$$

dividend – the number to be divided in a division problem

division – the operation of making equal groups to find the number in each group or to find the number of equal groups

divisor – the number by which another number is divided

doctorate – the highest post-graduate degree awarded by a university

dot plot – a graph that uses dots above a number line to show the frequency of data

E

edge – the line segment where two faces of a solid figure meet

equality symbol (=) – a symbol that indicates two quantities are equivalent

equation – a number sentence that uses the equal sign to show two expressions are equivalent

equilateral triangle – a triangle with three congruent sides

equivalent – the same in value or amount

estimate – *noun*: an answer that is close to the exact answer; *verb*: to guess about

evaluate – to find the value of an expression

exclusive – not contained in a set; for example, the set of integers between -3 and 1, exclusive, is {-2, -1, 0}

expense – an amount of money used to buy goods or services

exponent – the number of times a base is used as a factor; for example, 4 is the exponenet in 3^4

expression – a mathematical combination of numbers, operations, and/or variables

F

face – a flat surface of a 3-dimensional solid figure

factor – a number that is multiplied by another number to find a product

    The factors of 6 are 1, 2, 3, and 6.

factor tree – a diagram showing the prime factorization of a number

Federal Application for Student Aid (FAFSA) – free application used to determine eligibility for federal student aid

federal minimum wage – lowest hourly amount allowed by the federal government to be paid to a worker

financial aid – money given or loaned to assist in paying for college expenses

financial institution – a business that deals with money, deposits, investments, and loans, rather than goods or services

finite – having a limited number of values

formula – a general mathematical statement or rule

fraction – a number that names a part of a whole or part of a group

frequency – the number of times an event happens

frequency table – a table listing each value that appears in a data set followed by the number of times it appears

G

gap – the part of a data distribution where there is no data

grace period – a period of time after a payment is due before the borrower is charged fees or other penalties

grant – a type of financial aid, typically based on need, not requiring repayment

grouping symbols – braces, brackets, or parentheses used to group numbers, symbols, and/or variables

H

height – the distance from bottom to top

histogram – a graph that uses bars to show the frequency of data within equal intervals

horizontal – the direction from left to right; parallel to the horizon

hundredth – one of 100 equal parts; the second place to the right of the decimal point

I

identity property of addition – the property which states that the sum of any number and zero is equal to that number

$$a + 0 = a$$
$$8 + 0 = 8$$

identity property of multiplication – the property which states that the product of 1 and any factor is equal to the factor

$$1 \cdot b = b$$
$$1 \cdot 9 = 9$$

improper fraction – a fraction in which the numerator is greater than or equal to the denominator

inclusive – contained in a set; for example, the set of whole numbers from 3 to 7, inclusive, is {3, 4, 5, 6, 7}

income – the amount of money received in a particular time (weekly, monthly, yearly) in exchange for work

independent quantity – a quantity whose value determines that of a dependent quantity

x	1	2	3	4
y	8	9	10	11

independent quantities: {1, 2, 3, 4}

inequality – a mathematical sentence that shows the relationship between quantities that are not equal using the symbols <, >, ≤, ≥, or ≠

inequality symbol (<, >, ≤, ≥, ≠) – a symbol that compares two quantities

< means "less than"
> means "greater than"
≤ means "less than or equal to"
≥ means "greater than or equal to"
≠ means "not equal to"

infinite – having an unlimited number of values

input-output table – a table of values that follows a rule; used to show the pattern in x-values and their corresponding y-values

integers – the set of whole numbers and their opposites

interest – money paid by a borrower in exchange for using a lender's money for a certain period of time

interquartile range (IQR) – the difference between the upper quartile and the lower quartile

intersect – to meet or cross at a common point

inverse operations – opposite operations; for example, addition is the inverse operation of subtraction

inverse property of addition – the property which states the sum of a number and its opposite is zero

$$a + (-a) = 0$$
$$3 + (-3) = 0$$

inverse property of multiplication – the property which states the product of a number and its reciprocal is one

$$a \cdot \frac{1}{a} = 1$$
$$6 \cdot \frac{1}{6} = 1$$

interval – the distance or space between values; the set of points between two numbers

isosceles trapezoid – a trapezoid in which the non-parallel sides are congruent and the base angles are congruent

isosceles triangle – a triangle with two congruent sides

J

justify – to explain why a solution was chosen

L

late fee – a charge for not paying a bill by the due date

lender – a person or business that loans money

length – the distance from one end of an object to the other

line – a straight path that extends infinitely in opposite directions

linear angle pair – two adjacent angles that are also supplementary

loan – money that is borrowed by a person to purchase goods or services; the borrower agrees to repay the money over a set period of time with interest

lower quartile (Q_1) – the median of the lower 50% of the data set

lowest terms – the simplest form of a fraction in which the numerator and denominator have no common factor except 1

M

master's degree – a post-graduate degree typically earned after 3 years of study from a college or university

maximum value – the greatest value in a set of data

mean – the sum of the numbers in a set of data divided by the number of pieces of data; the average

measure – to find the size, weight/mass, or capacity of an item using a given unit

measures of center – a value at the center or middle of a data set; often represented using the mean or median

measures of spread – describes how the data is spread out; often represented using the range or interquartile range

median – the middle number in a set of data when the data are arranged in order

metric system of measurement – a measurement system used throughout the world, based on multiples of 10

minimum balance – the minimum amount a financial institution requires that a customer maintains in an account

minimum payment – the lowest amount that a borrower is required to pay each month toward a credit balance and remain in good standing

minimum value – the smallest value in a set of data

mixed number – a number composed of a whole number and a fraction; such as, $2\frac{3}{4}$ or $5\frac{1}{8}$

mode – the number or category that appears most frequently

model – a drawing, diagram, or smaller version of something that represents the actual object

multiple – the product of a given number and any whole number

Multiples of 5 include 5, 10, 15, 20, ...

multiplication – the operation using repeated addition of the same number; the combining of equal groups

multiplicative relationship – a relationship among a set of ordered pairs in which each *x*-value is multiplied by the same factor to find each corresponding *y*-value

The table shows a multiplicative relationship because each *x*-value is multiplied by 4 to find each *y*-value.

x	0	1	2	3
y	0	4	8	12

N

negative number – a number that is less than 0

number line – a line on which points correspond to numbers

number sentence – an equation or inequality that uses numbers and operation symbols

numerator – the top number in a fraction; the number of equal parts being considered

numerical data – values that can be counted or measured, such as dollars earned or minutes spent on homework

numerical equation – a number sentence that uses an equal sign to show that two numerical expressions are equal

numerical expression – a mathematical combination of numbers and operation symbols

O

obtuse angle – an angle measuring more than 90° but less than 180°

obtuse triangle – a triangle that has one angle greater than 90°

occupation – a job or profession

online banking – a system that allows customers of a financial institution to perform banking activities using the Internet

opposite numbers – two numbers that are the same distance from 0 on the number line but are on opposite sides of 0; 2 and -2 are opposite numbers

Opposite Side-Angle Theorem – if one side of a triangle is longer than another side, then the angle opposite the longer side is larger than the angle opposite the shorter side

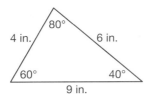

order of operations – the rules for the order of performing operations in expressions with more than one operation:
1) parentheses
2) exponents
3) multiplication/division from left to right
4) addition/subtraction from left to right

ordered pair – a pair of numbers, (x, y), used to locate a point on a coordinate plane

origin – the point on a coordinate plane where the x-axis and the y-axis intersect; (0, 0)

outlier – a data value that is widely separated from the other data values in the data set

overdrawn – the status of an account where the money taken out is more than what is available

overdraft – an arrangement between a financial institution and a customer in which the institution agrees to pay a debit on an overdrawn account

P

parallel – never meeting or intersecting; always the same distance apart

parallel lines – two lines in the same plane that never intersect

parallelogram – a quadrilateral with opposite sides that are parallel and congruent

parentheses – symbols used to group part of a mathematical expression or equation; ( )

partition – to divide or separate a whole into parts

pattern – a regularly-repeated arrangement of numbers, letters, shapes, etc.

peak – the maximum frequency of a data set

per – each or one

percent – the ratio of a number to 100

percent bar graph – a graph in which a bar represents a sum of 100%

perimeter – the distance around a two-dimensional figure

perpendicular – intersecting at right angles

perpendicular lines – lines that intersect to form right angles

plot – to determine and mark points on a coordinate plane or number line

point – an exact location or position; a point may be represented by a dot

polygon – a closed figure made of line segments

positive number – a number that is greater than 0

post-secondary education – refers to any education received at a community college, college, university, etc.

prime factorization – a number expressed as the product of its prime factors

prime number – a number that has exactly two positive factors, itself and 1

product – the answer to a multiplication problem

proportion – a mathematical sentence stating that two ratios are equal; for example, $\frac{2}{6} = \frac{6}{18}$

Q

quadrant – any of the four regions of a coordinate plane

quadrilateral – a polygon with four sides and four angles

quantity – an amount

quotient – the answer to a division problem

R

range – the difference between the highest and lowest values in a set of data

rate – a ratio in which two quantities with different units of measure are compared

ratio – a comparison of two quantities

rational number – a number that can be written as $\frac{a}{b}$ in which a and b are integers, but b is not equal to 0; $\frac{2}{3}, \frac{7}{4}, 1\frac{2}{5}$ are examples of rational numbers

ray – a part of a line that has one endpoint and extends forever in the other direction

reasonable – logical or sensible

reciprocal – one of two numbers whose product is 1; the reciprocal of 2 is $\frac{1}{2}$

rectangle – a parallelogram with four right angles

rectangular prism – a three-dimensional figure with six rectangular faces

regular polygon – a polygon in which all sides are the same length and all angles have the same measure

relationship – a connection or pattern found between numbers

relative frequency – a value determined by dividing the frequency by all possible outcomes

remainder – the number left over after dividing into equal groups

repeating decimal – a decimal in which one or more digits repeat indefinitely; $0.3333\ldots = 0.\overline{3}$

representations – ways to display a math concept; may be in the form of an algebraic expression or equation, a graph, a table, or a verbal description

rhombus – a parallelogram in which all four sides are congruent and opposite angles are congruent

right angle – an angle with a measure of 90°

right triangle – a triangle with one right angle

round – to approximate a number to a given place value

rule – a procedure that a pattern must follow

S

savings account – a bank account that allows a customer to deposit and withdraw money and earn interest from the bank

scalene triangle – a triangle with no congruent sides

scholarship – money awarded based on academic or other achievement, not requiring repayment

service charge (fee) – a fee charged by a financial institution for a service provided to customers

set – a collection of distinct elements

simplest form – the form of a fraction in which the greatest common factor of the numerator and denominator is 1

simplify – apply the properties of operations to an expression to make computation easier

skewed distribution – describes the shape of data that is pulled in one direction away from the center resulting in a tail on the left or right side

solution – any value for a variable that makes an equation or an inequality true

square – a special rectangle with 4 sides of equal measure

squared – raised to the power of 2; for example, 4 squared = 4^2

square unit – a square with a side length of one unit, used to measure area

statistical question – a question that anticipates variability in the answer

stem-and-leaf plot – a display of data in which digits with larger place values are named as stems and digits with smaller place values are named as leaves

strip diagram – a model or drawing that looks like a strip of tape; a strip diagram is used to represent number relationships

subset – a set within a set

supplementary angles – two angles whose measures sum to 180°

sum – the answer to an addition problem

survey – to ask many people the same questions in order to gather data

symmetric distribution – describes a set of data in which the shape of the left side is roughly the same shape as the right side

T

table – information organized in columns and rows

tenth – one of 10 equal parts; the first place to the right of the decimal point

term (in an expression) – a number, variable, or product of numbers and variables

terminating decimal – a decimal that has a definite number of digits after the decimal point; for example, $\frac{1}{5} = 0.20$

thousandth – one of 1,000 equal parts; the third place to the right of the decimal point

three-dimensional figure – a solid figure that has length, width, and height

transaction – an exchange or transfer of funds

transfer – the act of moving money from one place to another

trapezoid – a quadrilateral with exactly one pair of parallel sides

trend – a pattern in a set of data

triangle – a polygon with 3 sides and 3 angles

Triangle Inequality Theorem – the sum of the lengths of any two sides of a triangle is greater than the length of the third side

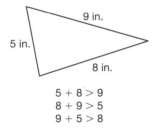

Triangle Sum Theorem – the sum of the measures of the angles in a triangle is 180°

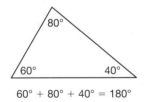

two-dimensional figure – a plane figure that has length and width

U

uniform distribution – describes the shape when the values of the data set are equally spread across the range

unit rate – a comparison of two measures with one term having a value of 1; may also be called unit price or unit cost

upper quartile (Q_3) – the median of the upper 50% of the data set

V

variability – measures how much data points differ from each other

variable – a letter or symbol used to represent a number

Venn diagram – a diagram showing relationships between sets

vertex/vertices – the point where two rays meet, where two sides of a polygon meet, or where the edges of a 3-dimensional figure meet; the top point of a cone or pyramid

vertical – straight up and down; perpendicular to the horizon

vocational training – training provided for a specific field or occupation

volume – the number of cubic units needed to fill the space occupied by a solid

W

wage – a regular payment, usually hourly or weekly, paid to a worker by an employer

weight – the heaviness of an object as determined by its mass

whole numbers – the numbers used to count and zero (0, 1, 2, 3, 4, ...)

width – the measure or distance across something from one side to the other

withdrawal – money removed from a savings or checking account

work-study – a student aid program that provides a part-time job while in school to help pay expenses

X

x-axis – the horizontal axis on a coordinate plane

x-coordinate – the first number in an ordered pair, locating a point on the x-axis of a coordinate plane

Y

y-axis – the vertical axis on a coordinate plane

y-coordinate – the second number in an ordered pair, locating a point on the y-axis of a coordinate plane

Z

zero pair – a number and its opposite; for example, 5 and -5

Notes

Notes

Notes

Notes

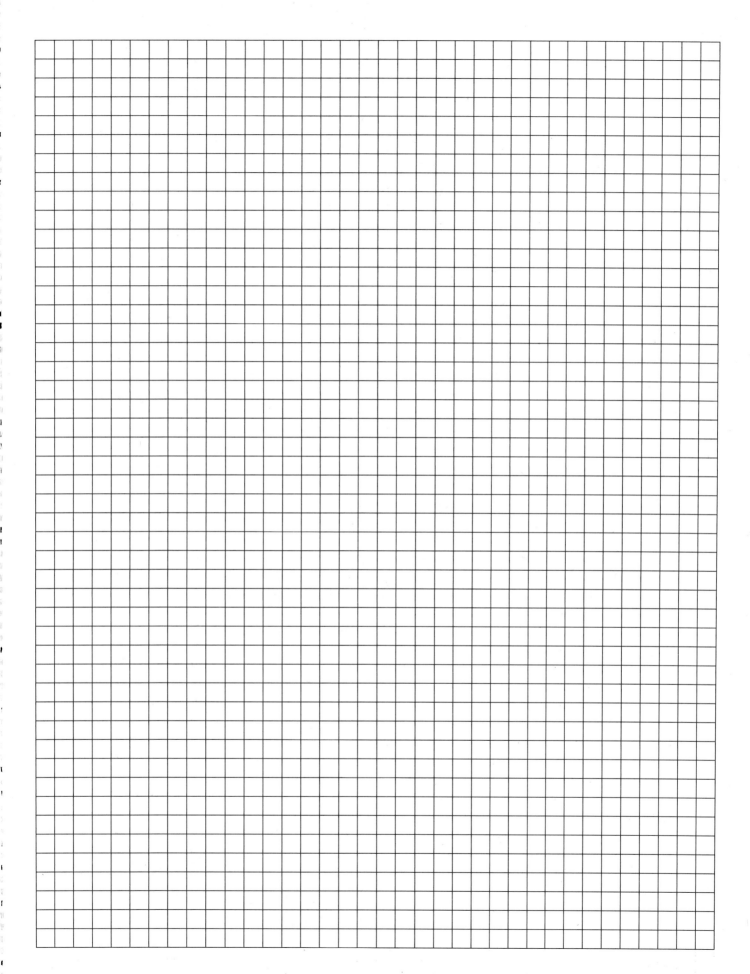

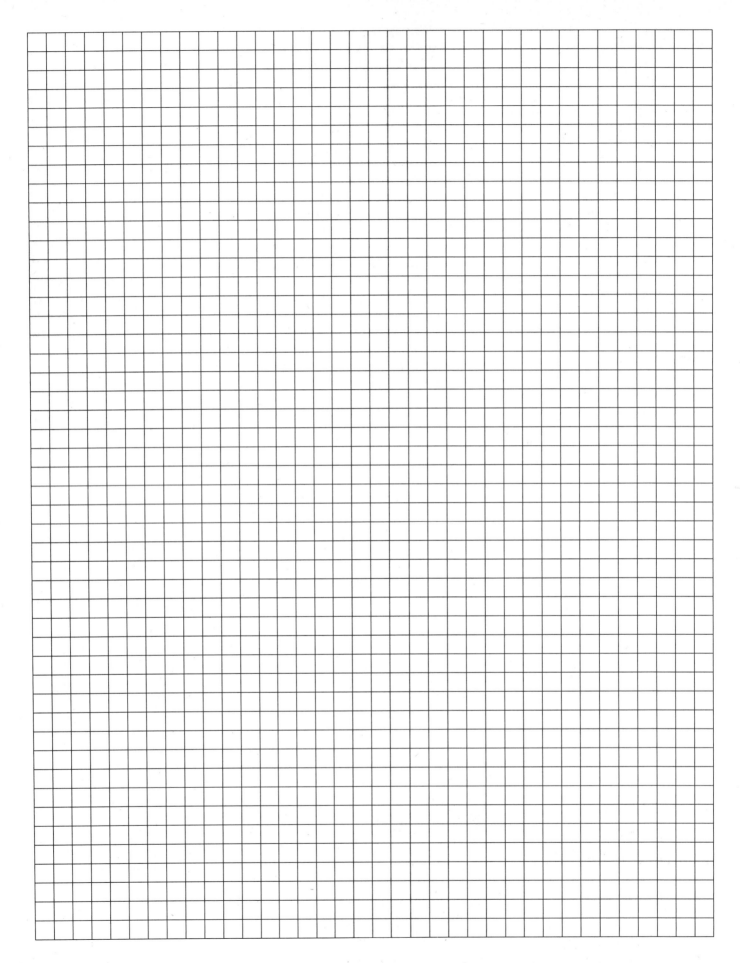

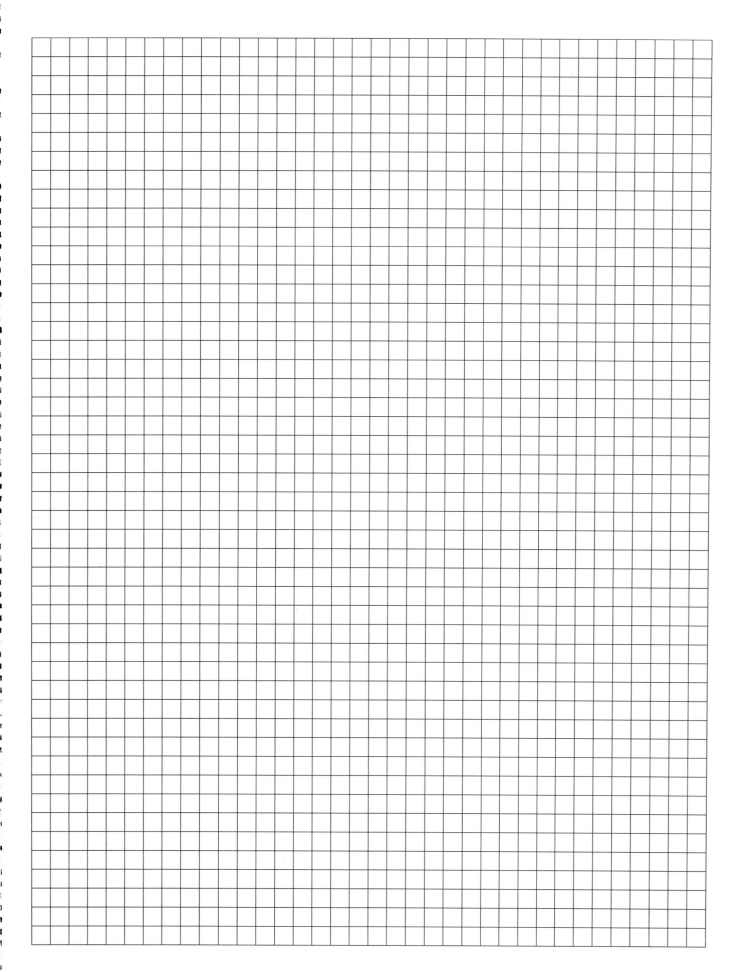

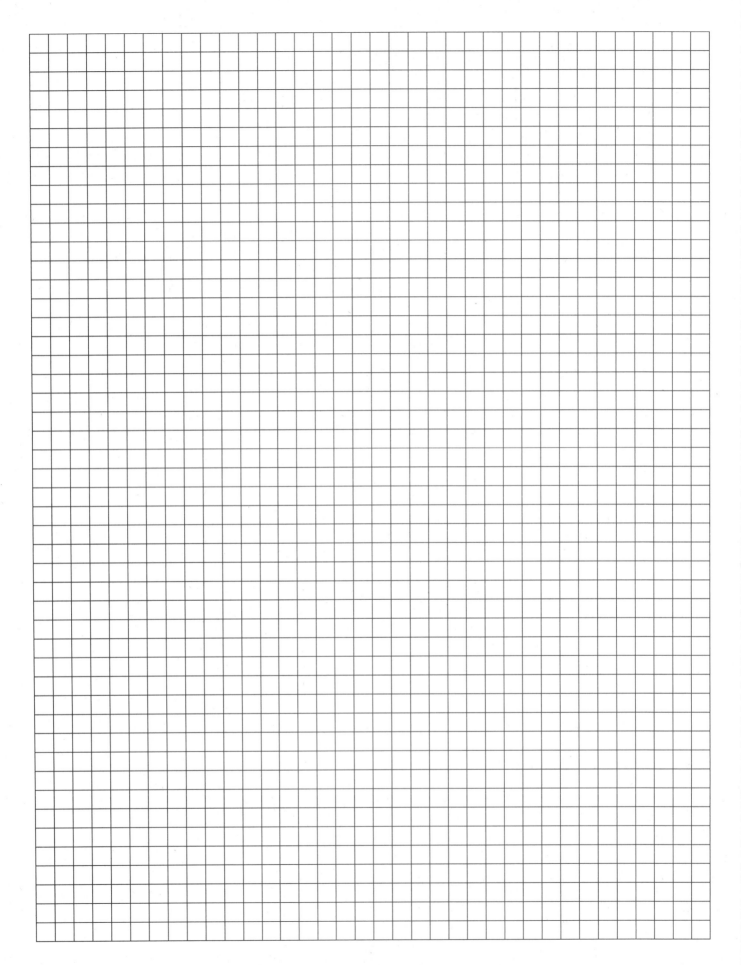

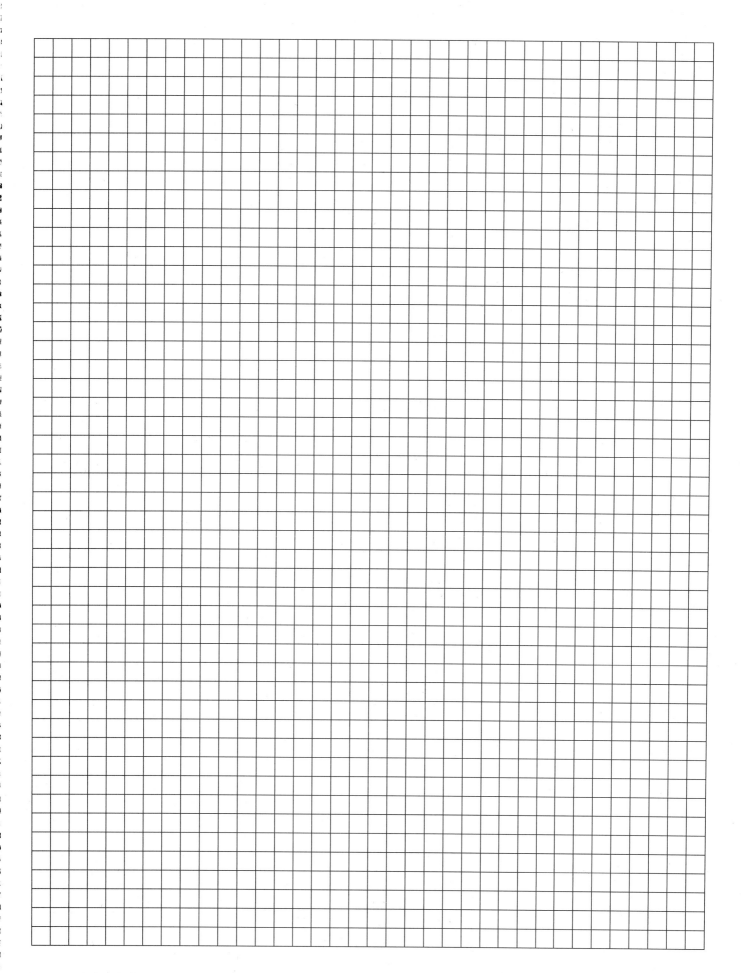

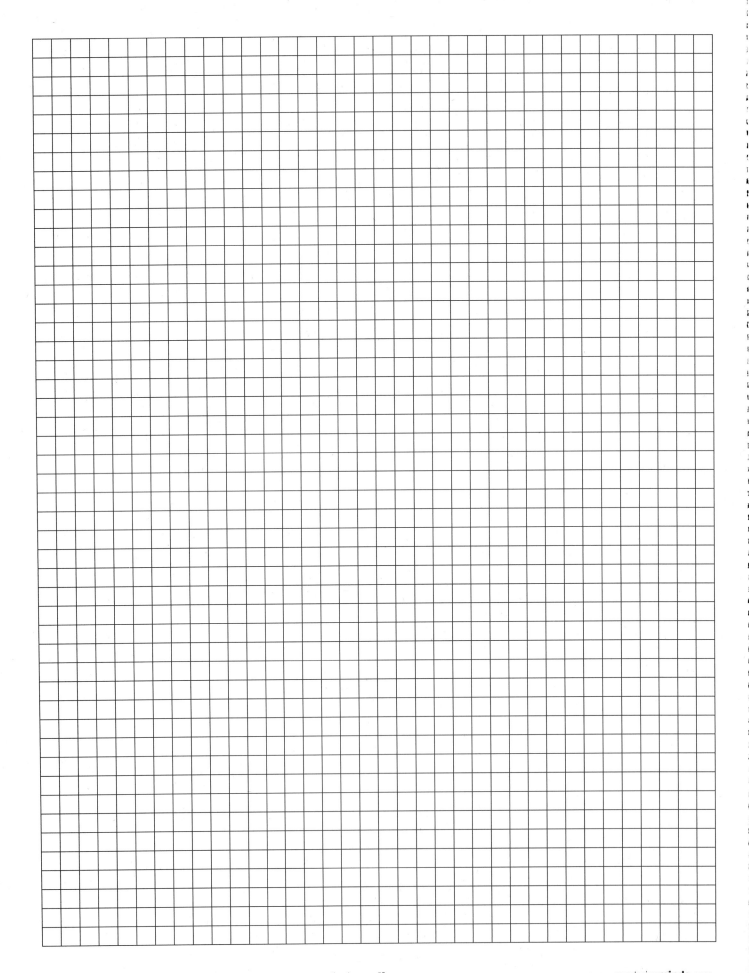

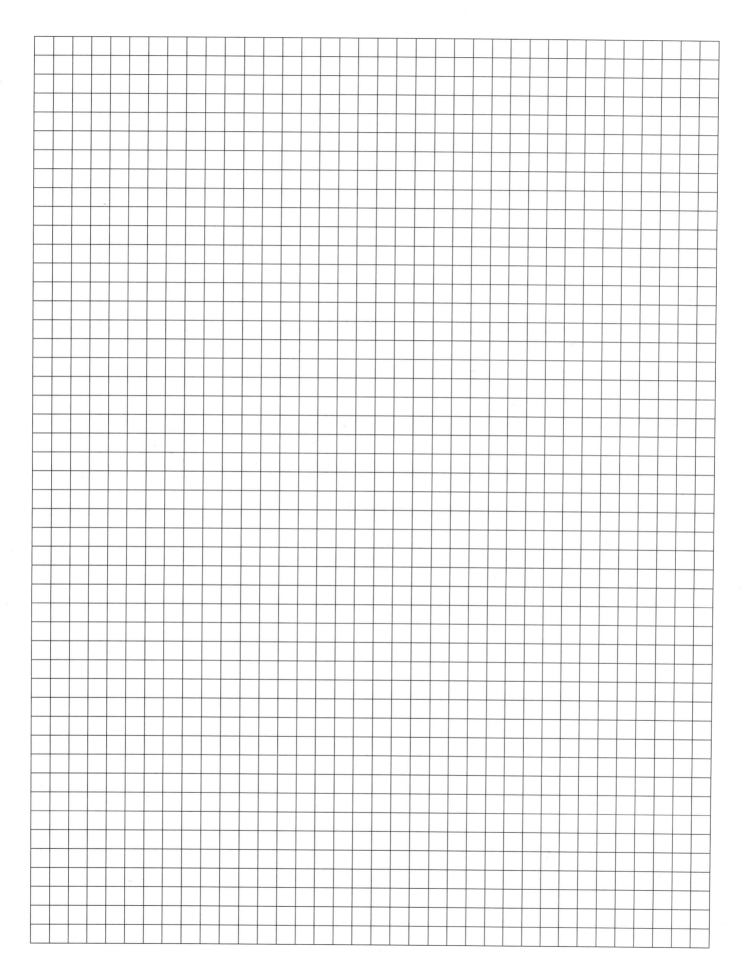

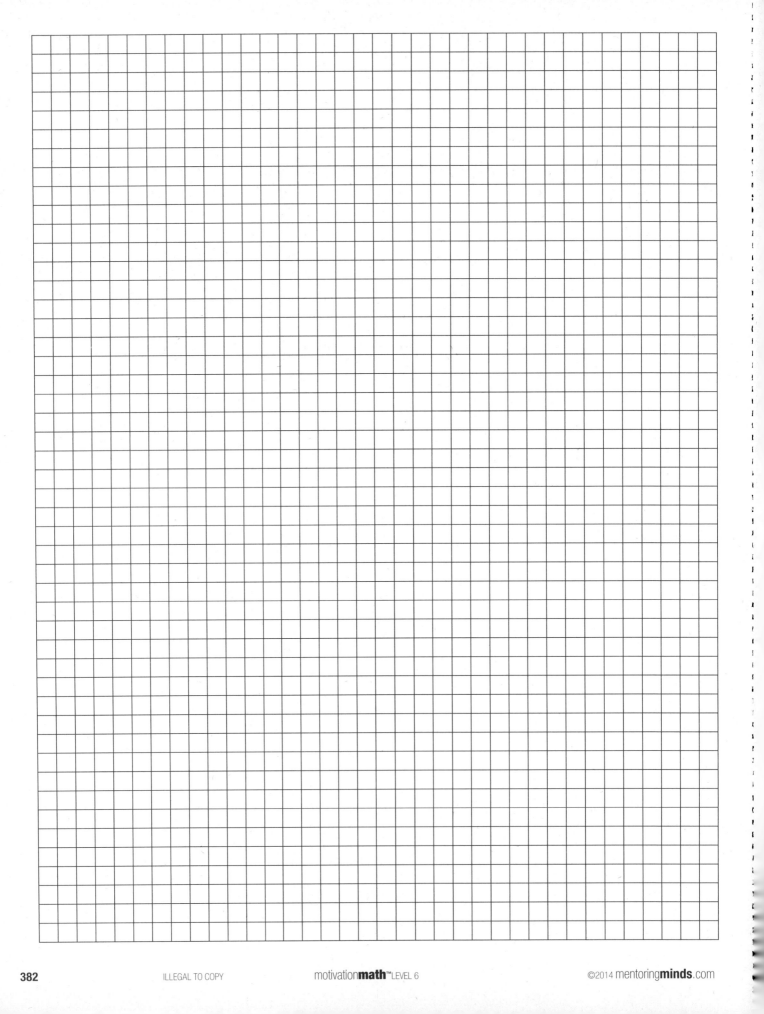

Grade 6 Mathematics Reference Materials

Area

Triangle	$A = \frac{1}{2}bh$
Rectangle or parallelogram	$A = bh$
Trapezoid	$A = \frac{1}{2}(b_1 + b_2)h$

Volume

Rectangular prism	$V = Bh$

Grade 6 Mathematics Reference Materials

LENGTH

Customary	Metric
1 mile (mi) = 1,760 yards (yd)	1 kilometer (km) = 1,000 meters (m)
1 yard (yd) = 3 feet (ft)	1 meter (m) = 100 centimeters (cm)
1 foot (ft) = 12 inches (in.)	1 centimeter (cm) = 10 millimeters (mm)

VOLUME AND CAPACITY

Customary	Metric
1 gallon (gal) = 4 quarts (qt)	1 liter (L) = 1,000 milliliters (mL)
1 quart (qt) = 2 pints (pt)	
1 pint (pt) = 2 cups (c)	
1 cup (c) = 8 fluid ounces (fl oz)	

WEIGHT AND MASS

Customary	Metric
1 ton (T) = 2,000 pounds (lb)	1 kilogram (kg) = 1,000 grams (g)
1 pound (lb) = 16 ounces (oz)	1 gram (g) = 1,000 milligrams (mg)